Guides to Professional

Series Editor:
Adrian Wallwork
Pisa, Italy

For further volumes:
http://www.springer.com/series/13345

Adrian Wallwork

Meetings, Negotiations, and Socializing

A Guide to Professional English

 Springer

Adrian Wallwork
Pisa
Italy

ISBN 978-1-4939-0631-4   ISBN 978-1-4939-0632-1 (eBook)
DOI 10.1007/978-1-4939-0632-1
Springer New York Heidelberg Dordrecht London

Library of Congress Control Number: 2014939422

Printed on acid-free paper

Springer is part of Springer Science+Business Media (www.springer.com)

Who is this book for?

This book is a guide to taking part in business meetings and negotiations, and to the socializing required before and after such events.

The book is aimed at non-native English speakers, with an intermediate level and above.

I hope that other trainers like myself in Business English will also find the book a source of useful ideas to pass on to students.

If you work in Academia, a better option for you is to consult Parts 3-6 of my book *English for Academic Correspondence and Socializing* (Springer Science), where some of the subsections in this book are taken from or adapted from.

How is this book organized?

Four main topics are covered

1. Meetings (Chapters 1 - 6)

2. Negotiations (Chapters 7 - 9)

3. Socializing (Chapters 10 - 12)

4. Understanding native English speakers (Chapters 13 - 14)

Within each chapter there are various subsections (see Contents for details), most of which are divided into FAQs (frequently asked questions). The book concludes with a chapter of useful phrases.

This book is designed to be like a manual or user guide – you don't need to read it starting from page 1. Like a manual it has lots of short subsections, and is divided into short paragraphs with many bullet points. This is to help you find what you want quickly and also to assimilate the information as rapidly and as effectively as possible.

You may find that occasionally the same concept is explained more than once but in different sections. This is because the book is not designed to be read from cover to cover, and some concepts may apply, for instance, both to participating in a meeting and in a social situation.

How should I use the table of contents?

The table of contents lists each subsection contained within a chapter. You can use the titles of these subsections not only to find what you want but also as a summary for each chapter.

Other books in this series

There are currently five other books in this *Guides to Professional English* series.

CVs, Resumes, and LinkedIn
http://www.springer.com/978-1-4939-0646-8/

Email and Commercial Correspondence
http://www.springer.com/978-1-4939-0634-5/

User Guides, Manuals, and Technical Writing
http://www.springer.com/978-1-4939-0640-6/

Presentations, Demos, and Training Sessions
http://www.springer.com/978-1-4939-0643-7/

Telephone and Helpdesk Skills
http://www.springer.com/978-1-4939-0637-6/

All the above books are intended for people working in industry rather than academia. The only exception is *CVs, Resumes, Cover Letters and LinkedIn*, which is aimed at both people in industry and academia.

There is also a parallel series of books covering similar skills for those in academia:

English for Presentations at International Conferences
http://www.springer.com/978-1-4419-6590-5/

English for Writing Research Papers
http://www.springer.com/978-1-4419-7921-6/

English for Academic Correspondence and Socializing
http://www.springer.com/978-1-4419-9400-4/

English for Research: Usage, Style, and Grammar
http://www.springer.com/978-1-4614-1592-3/

INTRODUCTION FOR THE TEACHER / TRAINER

Teaching Business English

I had two main targets when writing this book:

- non-native speakers (business, sales technical)
- Business English teachers and trainers

My teaching career initially started in general English but I soon moved into Business English, which I found was much more focused and where I could quickly see real results. The strategies I teach are almost totally language-independent, and many of my 'students' follow my guidelines even when speaking and writing in their own language. I am sure you will have found the same in your lessons too.

Typically, my lessons cover how to:

1. participate in meetings
2. write emails
3. make presentations and demos
4. make phone calls
5. socialize

This book is a personal collection of ideas picked up over the last 25 years. It is not intended as a course book, there are plenty of these already. It is more like a reference manual.

How to teach meetings skills

Teaching students how to take part in meetings is a lot of fun. Basically any discussion that you might normally have in your lesson can be transformed into a meeting. The language functions used in a discussion are essentially the same as those used in a meeting:

- expressing opinions
- agreeing and disagreeing
- interrupting and counteracting
- making suggestions
- clarifying misunderstandings
- drawing conclusions

The only difference is that some meetings tend to be more formal, may have a time limit and in some cases may have a chairperson.

As in a classroom discussion, you need to lay out some basic ground rules to make the discussion / meeting effective. The most important is that no one is allowed to dominate the conversation. Those students who are more extrovert should be encouraged to involve the more reticent students.

You can pre-teach all the useful phrases they will need in order to interrupt each other, make suggestions, clarify positions etc. You can even make it competitive by giving each student five useful phrases that they have to use in a natural way during the meeting. The first student to use all five is the 'winner'. However, bear in mind, that it is perfectly possible to have a meeting only using very few of the useful phrases listed in Chapter 15. But it is worth students learning them - even if they don't use them, they will at least be able to understand them when they hear them.

To learn more about conducting effective discussions and for ideas on what to talk about, see my books for Cambridge University Press: *Discussions AZ Intermediate* and *Discussions AZ Advanced*. These two books were not written specifically for a business audience, but given that the aim of your lessons will be to teach students how to participate in a discussion, the actual topic of the discussion does not necessarily have to be business related.

See also my publications section on LinkedIn, where you will find more books on discussions that you can buy on iTunes and Amazon.

How to teach negotiating skills

Basically we negotiate every day of our lives with bosses, colleagues, children, family members etc. A negotiation, in EFL terms, is not much different from a meeting, except that it will contain a lot more conditional sentences (first and second forms), e.g.

If you *lower* your price, we *will increase* our order.

If you *lowered* your price, then we *would increase* our order.

So you need to make sure that your students have mastered the difference between these two conditional forms. Then you need to set up discussions based on hypotheses such as what if: your boss was a woman rather than a man (or vice versa), your company was bought by an American company, you were forced to take a pay cut, your company relocated to China etc. There are endless hypothetical situations you can think of. You can even get your students to practise writing second conditional sentences containing interesting work-related scenarios.

To get more ideas, see the chapter on *Negotiations* in my book *Business Options* (OUP).

How to teach socializing skills

I find that many students, especially technical experts, have difficulties socializing effectively even in their own language. Typical issues are:

- talking too much and dominating the conversation (this is made even worse when talking in English, as they have more control over what is being said)
- not asking questions (may be due the fact that questions are quite difficult to form in English, but more often due to a lack of curiosity)
- being silent (i.e. as a result of poor English, or being too shy and embarrassed)

Chapter 10 should help you to help your students on the above issues.

Brainstorm your students on typical situations where they have to use their socializing skills. Then for each scenario build up dialogs that incorporate the strategies outlined in Chapter 11.

Also, you really need to work on stimulating your students curiosity about other people and places.

There are also many skills books available from the major EFL publishers (OUP, CUP, Longman and Macmillan) which cover meetings, negotiations and socializing. I suggest you dip into these books rather than following them exercise by exercise. Just pick out the exercises that you think students will find most interesting.

If you work in-house, get involved with the company / companies where you teach. You will find your work much more satisfying!

Contents

1.1 How should I email an existing client / supplier to arrange a meeting?

When emailing people you already know, you can follow this structure:

1. announce that you want to arrange a meeting (possibly also suggesting the time)

2. explain the need for the meeting

3. say who should attend and why

4. state the time and place of the meeting (if not mentioned earlier)

5. say whether you are attaching an agenda in the current email, or whether you will be sending one later (if so, when)

6. reiterate the importance of the meeting

7. ask for confirmation of attendance

Below is an example email to a group of people who already know each other:

Dear all,

Can we arrange a conference call for 15.00 on Monday 21 October?

I would like us to discuss the xyz project.

The following people would benefit from attending as they will be an integral part of the project: Yohannes, Sergei, Brigitte and Wei.

I will be sending you an agenda within the next two days, and I would appreciate your feedback on this agenda by the end of next week.

I would like to stress that this call should help us to move forward our plans for the project and ensure its success.

Please could you confirm your attendance by tomorrow evening.

Thank you.

A. Wallwork, *Meetings, Negotiations, and Socializing,*
Guides to Professional English, DOI 10.1007/978-1-4939-0632-1_1,
© Springer Science+Business Media New York 2014

1.2 What's the best way to email a potential client to arrange a meeting?

When emailing people you do not already know, you can follow this structure:

1. introduce yourself / explain how you got the recipient's email address

2. announce that you want a meeting

3. explain the purpose of the meeting

4. suggest a possible date / time

5. motivate the recipient to agree to the meeting

Below is an example email to a person that the sender does not know or does not know very well:

> Your name was given to me by Sergei Kapkov who though you might be interested in ...
> ... I am the sales manager at .... and as you can see from our website (www.xxx.yyy), we specialize in ...

> Next month I plan to be in your are area and I was wondering whether you might find the time to meet, so that I can give your more details about ...

> The meeting should not take up more than 30 minutes of your valuable time. Please let me know whether you might be free in the first or second weeks of next month.

> From what Mr Kapkov has told me, I am sure we could save you up to 25 % in your budget for ...

> I look forward to hearing from you.

1.3 I want to set up a 1-1 meeting with someone I already know. What email should I write?

Emails to people who you already know or who work for the same company tend to be much shorter and less formal. The email below is to a colleague who the sender already knows.

> I'm in London w/c 30 August so this might be a good time to meet up and discuss anything you might have in mind regarding …

> Let me know when would be a good time for you, though my preference would be for early afternoon so that I can get the late afternoon flight back.

> Best

The next email is from a person in one branch of a company who is trying to find out who is the appropriate person to talk to in another branch of the same company.

> Next week I will be in NY and I would like to meet to discuss our recruitment process. Are you the right person to be talking directly to the recruitment agencies and give them any details they might need?

> If you could get back to me by the end of today that would be great.

> Best regards

1.4 How can I arrange a meeting via telephone?

It may be quicker to set up a meeting by contacting the person directly by telephone, then the decision can be made immediately.

The first dialog below takes place between two people who do not know each other. In reality the meeting arrangement is unlikely to be achieved in just one phone call but would probably be concluded over a series of phone calls and emails.

Good morning. My name is *name* and I am responsible for *position* at *name of company*. I found your name on your website and I was wondering whether you would be the right person to talk about ...

Possibly yes, can you give me a few more details?

Well, we work in the field of ... and we have just developed a ... and I would very much like to have the opportunity of meeting you and giving you a demo.

Unfortunately I am extremely busy at the moment.

I understand, but I guarantee I would need no more than 20 minutes of your time. When might you be free later this month or early next month.

I'm busy right through to the end of this month. Maybe the second week of next month.

Perfect. Would Thursday the 13th suit you?

Actually, no I have meetings all day on the 13th.

What about the previous day, Wednesday the 12th?

That would be fine.

OK. Well thank you very much for your time. I will send you an email detailing the day and time, and I will also give you a little more information about the product and how I believe it could save your staff not only a considerable amount of time but also money.

Sorry, can you just repeat your name for me.

Yes, it's *name*. You can expect my email within the next hour.

OK. Goodbye for now.

Goodbye and thank you.

1.4 How can I arrange a meeting via telephone? (cont.)

In the above dialog note how the caller (in normal script):

- always tries to accommodate the prospect's requests
- reassures the prospect that the meeting will not take long
- does not insist too much but gets the meeting
- summarizes what has been decided
- promises a follow-up email
- does not talk to much and is concise in what he says

The next dialog is between colleagues who already know each other.

Hi, this is Petra from the Berlin office calling.

Hi Petra, how are you?

Fine thanks. You?

Great. [What can I do for you?].

The reason I am calling is that I would like to arrange a meeting to discuss …

OK. [Did you have a particular time and day in mind?]

Well, I was thinking next Monday at 4 o'clock, at your office. I am flying over for the day.

OK, but it would have to be short. I've got another meeting at 5.

Well it won't take more than 30 minutes, 45 at the most.

OK, that sounds fine. So next Monday at 4 o'clock in my office.

Perfect thanks.

For more details on how to make telephone calls, see the companion volume: *Telephone and Helpdesk Skills.*

1.5   I need to cancel / change the time of my meeting. How can I do this politely?

Follow this structure (both for email and phone).

- apologize and say you need to cancel / change time
- explain reason
- suggest alternative
- apologize again

For example:

> Hi, I am very sorry but I need to cancel our meeting (arranged for next Wednesday at 10.00). Unfortunately, something unexpected has come up that requires my urgent attention. Would it be possible for you on Thursday instead, at 10.00? Once again, I apologize for having to change our arrangements and I hope this does not cause you any problems.

2 PREPARING FOR A MEETING

2.1 How should I prepare for the meeting?

There are several factors that can contribute to a successful meeting.

- Find out as much as you can about the meeting before you go: a) topic b) who will be present (nationality, position in company, age)

- Decide exactly what it is you want to discuss, then note down any key words and phrases in English that you might need.

- Prepare a script of anything particularly important that you want / need to say. Then practise reading your script aloud. Modify to make it more concise and convincing.

- Try to predict what people are likely to say. Write down some key phrases that will help you to agree with or counter what they might say.

- Are the participants likely to agree with what you are going to say? If not, think of ways in English to counteract their objections.

2.2 How can I increase the chances of the meeting being useful for me?

In addition to preparing for the meeting (see 2.1 above), there are other things that you can do to ensure that you get the full benefits of the meeting.

- Try to sit near to the people who are likely to talk the most, this should enable you to hear better.

- If you need time to reflect on what is being said, suggest having a coffee break to enable you to collect your thoughts and prepare what you want to say.

- After the meeting, send the chairperson an email summarizing what you think has been the outcome of the meeting.

A. Wallwork, *Meetings, Negotiations, and Socializing*,
Guides to Professional English, DOI 10.1007/978-1-4939-0632-1_2,
© Springer Science+Business Media New York 2014

2.3 I will be chair of the meeting. What guidelines should I follow?

When chairing a meeting try to do the following:

- Define a clear purpose for the meeting
- Follow an agenda
- Start and finish meetings on time
- Increase participant involvement during meetings
- As far as possible, reduce dysfunctional practices during meetings (e.g., side conversations, discussion of irrelevant issues, hidden agendas)
- Resolve emotional conflicts among meeting attendees in a professional manner

2.4 What is brainstorming? How should it be conducted?

The objective of a brainstorming session is to encourage participants to come up with new and creative solutions to a problem. The difference from an ordinary meeting is that

- in the initial stages there are no assumptions on what constitutes a good solution, so no idea is criticised or evaluated
- participants should have no constraints on generating ideas, even ideas that initially may seem irrelevant or wildly impractical
- ideas are merely used to stimulate or develop another idea (related or not)

If you are chairing a brainstorming session, ensure that:

- you establish the duration of the session
- there is a clear focus – the ideas can be widely divergent, but they should all relate to solving a specific problem
- quiet or reticent attendees also participate actively
- someone keeps a note of the ideas on a whiteboard or flipchart

When the session has reached its set duration, now is the time to evaluate and discuss the ideas.

3 USING ENGLISH BEFORE, DURING, AND AFTER A MEETING OR NEGOTIATION

3.1 What are native speakers likely to think about my English?

Many business people who are native speakers of English do not speak any other language apart from English. Moreover, they may have previously had little contact with non-native speakers and thus may not be familiar with the difficulties that you have expressing yourself in English.

From a native English speaker's point of view:

* your grammar is relatively unimportant

* it is better if you are reasonably fluent and inaccurate (but not too much!), than slow and totally accurate

* irrespective of your position in the company hierarchy, if you speak the best English you are more likely to be addressed than your colleagues

* you need to know the typical phrases used in a meeting (see Section 15) and of course the key vocabulary of your field

These points do not imply that your English has to be perfect. Instead, you should have a good command of just those things that you need in order to be able to carry out your job.

3.2 What typical phrases will I need during the meeting or negotiation?

Meetings, especially those conducted by conference calls (see Chapter 6) are a difficult form of communication for non-native speakers. However, if you prepare well, and learn some useful phrases to help you out of difficulty, then these meetings need not be such a terrifying experience!

The typical phrases are listed in Chapter 15. Such phrases tend to be used very frequently in meetings. Even if you don't use them yourself, if you familiarize yourself with them you will be able to recognize them when others use them.

A. Wallwork, *Meetings, Negotiations, and Socializing,*
Guides to Professional English, DOI 10.1007/978-1-4939-0632-1_3,
© Springer Science+Business Media New York 2014

3.3 How important is it for me to see the agenda before the meeting?

If you have not already received it, ask to be emailed the agenda of the meeting. This will enable you to have time to write down any questions you would like to ask. Given the construction of questions in English (inversion of subject and verb, use of auxiliaries) formulating questions when you are speaking can be quite tricky. if you write the questions down beforehand, you will:

- have the time to formulate them correctly
- be able to refer to them during the meeting
- have a chance to think about how the attendees will react to your questions, and then think of possible answers

Likewise, you can try and predict the questions that you might be asked. Again, you can prepare your answers in advance.

By doing this preparatory work, you will feel more confident before the meeting (either face to face or remotely) begins and this will improve your performance and enhance your company's image.

3.4 Is it a good idea to arrive early to the meeting?

Not everyone will arrive at the meeting at the same time. It is a good idea to arrive early. You will then have the opportunity to meet and chat with the other attendees as they arrive. In the case of people you have never met before this will give you a chance to:

- speak some English in a non-critical situation, i.e. it will not be a problem if you make some mistakes in your English during small talk with other participants
- have an opportunity to hear the voices of the other participants and to tune in to the way they speak

The result will be that you will be less nervous when the meeting starts and you will stand a greater chance of understanding what the other attendees are saying.

See the Useful Phrases section to find out what to say to:

- other attendees who you have not met before
- people you have met before (but a long time ago, or only superficially)
- people who you know well

3.5   What can I do if the native speakers or fluent speakers dominate the meeting?

Unfortunately, whether deliberately or not, the native speakers (or the most fluent speakers) may tend to dominate a meeting because of their advanced linguistic skills. This can even lead to unwanted outcomes for the non-native speaking party.

It is critical that you make the native speakers aware of your lower level of English. You can do this in various ways:

1) announcing the fact that your English level is low:

> I am sorry, but my English is not great. Please could you speak slowly and make frequent summaries.

2) apologizing for the fact that you may need to make frequent interruptions

> I would like to apologize in advance if I need to interrupt you to clarify that I have understood.

3) suggesting frequent breaks – such breaks will not only enable you to rest your brain, but also will be an opportunity for you to evaluate what has been said so far and also discuss it with your colleagues

> Would it be alright to schedule short breaks every 30 minutes. This is because it is very tiring talking for long periods in a foreign language.

4) saying that you and your colleagues may need to discuss things in your own language

> I hope you don't think it rude if my colleagues and I occasionally say something in our own language.

By referring to one or more of the four points above you will be partially able to compensate for your linguistic disadvantage. Also, it is my belief that native speakers should be aware of the difficulties that non-natives may have in expressing themselves in English under stressful conditions.

3.6 My English is not great. How can I optimize it for the purposes of the meeting?

Below are some general rules about speaking in English during a meeting.

- Make the other members of the meeting aware of your level of knowledge of English, if low, before the meeting starts – this is essential (see 3.5).

- Don't hesitate to ask for repetitions or for the person to speak more slowly (provided that you have made them aware of your level)- the English speaker must remember that the only reason you can't understand is that he / she speaks English and not your language.

- Don't be afraid to interrupt – make sure you participate actively.

- Don't worry too much about grammatical mistakes. It is infinitely preferable to speak fluently and coherently with a lively voice, than with perfectly constructed sentences said in a slow monotone. Try to sound confident even if you aren't.

- Try to improve your intonation. Learn how to show enthusiasm or disapproval. Depending on your mother tongue, your usual intonation might seem rather rude in English or disinterested. If you can't help your intonation, at least make sure your facial expression reflects what you're trying to say.

- Try at least to pronounce the words in your particular field of business correctly with the right stress.

3.7   How should I use stress and intonation?

When participating in a meeting or a negotiation, it is essential that you make your point clearly. One way to do achieve this is to stress the key words.

Stressing a word means giving it more emphasis than the surrounding words. You stress a word by saying it a little bit louder and longer than the other words.

Most often the stress words are verbs and nouns, as these generally carry the most meaning. This is illustrated in the example phrases below, in which the chairperson of a meeting is checking the opinions of the participants. The words to stress are in italics

Do you all *agree* on that?

Does anyone have any *comments*?

What are your feelings about the *budget*?

What are your *views* on this?

If you put the stress on words (pronouns) such as *you, my, his, she*, or on names of people, rather than on the verb or noun, this indicates that you are trying to contrast the views of two (or more) different people.

I understand what *you're* saying, but I am not clear what *Martin's* point is.

I don't think that's for *us* to decide, surely it's for *them* to decide.

Sorry, I meant *your* project not *her* project.

In other cases, if you put the stress on the pronoun you might confuse your listener. For instance compare:

That's not what Martin *meant*.

That's not what *Martin* meant.

In the first case, the speaker is saying that we have not understand the meaning of what Martin was trying to tell us. In the second case, the speaker is making a contrast between what Martin meant and what some other person meant.

3.7   How should I use stress and intonation? (cont.)

Sometimes you will need to stress the adjective or adverb

> You may be *right*, but *personally* I ...
>
> I'm not sure whether that's *feasible*.
>
> I don't want to sound *discouraging* but ...
>
> Am I making myself *clear*?
>
> This needs to be done *efficiently*.

To express what you think is the right thing to do, or to make a proposal sound more tentative, then you would probably want to stress modal verbs such as *must, should, may,* and *might*.

> It *might* be a good idea to contact them directly.
>
> Yes, we *should* move forward as fast as possible.
>
> This *must* be done before the end of the week.
>
> You *may* be right, but I think there's a strong possibility that ...

If you want to give special emphasis to a verb, you can place the auxiliary (*do, does*) before it, and stress that auxiliary. In the examples below the speaker is underlining the fact that she has understood and appreciated the other person's point of view, but that she has some reservations.

> I *do* understand what you are saying, but ...
>
> What you are saying *does* make sense, however ...

Very occasionally, you will need to stress a preposition. However, with prepositions the stress is normally very slight.

> Are you *with* me? (= are you following what I am saying?)
>
> Can we do that *before* the break, rather than *after* the break?

In summary, stressing particular words in a sentence:

- helps the listener understand the key points of what you are saying
- helps you to make differentiations (e.g. between different people, different approaches)
- stops your speech from being monotonous

3.8 Should I follow up the meeting with an email?

It is always standard practise to follow up a meeting with an email or other document that summarizes and confirms what was agreed during the meeting.

Below is a typical email that the chairperson might send after a meeting.

Dear All

Thank you all for making yesterday's meeting a success.

Attached are the minutes of the meeting. The main points we are agreed on are:

1) x

2) y

3) z

We scheduled the next meeting for Monday 6 October at 10.00 am. Please can you confirm by the end of tomorrow that you will be able to participate.

Please feel free to contact me if you have any comments on the minutes or on what was agreed.

Best regards

If the chairperson does not send you such a summary or if the meeting was informal with just a few attendees, then it is not a bad idea to make the summary yourself. below is an example email:

Hi

It was good to see you yesterday.

I just wanted to make sure that I had understood everything correctly and to summarize what we agreed.

Firstly, we decided that …

Secondly, the project is now going to be under the responsibility of …

Thirdly, we have scheduled monthly meetings for the first Monday of every month.

There are two points I am not entirely clear about:

1) x?

2) y?

I would be grateful you could get back to me today with your answers to my two queries.

Thanks

4 HOW TO MEET KEY PEOPLE AT NETWORKING EVENTS AND TRADE FAIRS

4.1 How can I improve my chances of meeting key people at a networking event or trade fair?

Networking is much simpler if you have a clear idea in advance of who you would like to meet (hereafter 'your key person'). A simple way to do this is to:

- look at the conference / trade fair program and find the names of key persons

- find information about them from their personal pages or company website

- find a photograph of them so you will be able to identify them in a room from a distance

Then you need to prepare questions in English that you wish to ask them.

You should also predict how they might answer your questions. This will increase your chances of understanding their answers and will also enable you to think of follow-up questions.

A. Wallwork, *Meetings, Negotiations, and Socializing,*
Guides to Professional English, DOI 10.1007/978-1-4939-0632-1_4,
© Springer Science+Business Media New York 2014

4.2 Before the event, should I email people I want to meet?

You will massively increase your chances of having a conversation with your key person if you email them beforehand. For example, if you are going to a trade fair you can email them to say that you would be interested in meeting them:

Subject: XYZ Trade Fair: meeting to discuss ABC

Dear Milos Dubrakov

I see from the website of the XYZ Trade Fair that your company will be attending the event. I am a sales manager at Boskov Electronics, which produces ABC components that could be easily integrated into your system.

I was wondering if you might be able to pass by our stand (Stand 127, next to the Apple stand) where I could show you some of our products (www.boskovelectronics.com / products). This wouldn't take up more than ten minutes of your time.

Alternatively, I could come by your stand. Please let me know a day and time that would suit you. I am free throughout the event, apart from on the Monday afternoon.

I look forward to hearing from you.

The structure is as follows:

1. say how you know about the key person (i.e. they are attending the same trade fair as you)

2. briefly describe what you do

3. show how what you do relates to what they do

4. indicate how long the meeting might last (keep it as short as possible)

5. suggest a possible meeting place and time, but show flexibility

Of course, there is no guarantee they will even open your email, but if they do you will have created an opportunity for a meeting. Such an email requires minimal effort. It also helps to avoid the embarrassment of having to walk up to a complete stranger and introduce yourself in English.

4.3   How can I motivate someone to meet me?

Although we sometimes do things purely for altruistic reasons, we are generally more motivated to help people if it seems that there might also be some benefit for us. It is thus a good idea to think of how a collaboration with you could benefit your key person—what knowledge do you have that would be useful for them, in what ways could you enhance their business, what contacts do you have that might be useful for them too?

4.4   How can I find out about someone in a discreet way?

If you think that the meeting you have arranged could significantly promote your business, then you need to do everything you can to ensure a successful outcome. Find out all you can about the person—find them on LinkedIn or Facebook, locate their company's website. Find out what is important both for them on a personal level and for their company. Find out what they are interested in aside from their work. Look for things that you might have in common.

> I read on your LinkedIn page that you previously worked for …
>
> I was looking at your profile on your company's website and saw that …
>
> A colleague mentioned that your company is investigating whether to …

However, although most people will not mind if you have investigated a little about their professional life, they may find it creepy (i.e. weird and disturbing) if you have been looking at their holiday photos on Facebook and know all about their hobbies. So be extremely careful how you refer to the things that you have learned about the person.

You can make your meeting much more beneficial if you are determined to find any person that you meet interesting. This will make you yourself more animated and thus appear more interesting to your interlocutor. You will also be less distracted as you will be focusing totally on the other person.

During the conversation restate and / or summarize the key points to check that you have understood. This is also a way to keep your mind alert and at the same time proves your appreciation of your interlocutor's remarks.

4.5 How should I introduce myself face-to-face to someone I have never contacted before?

First you probably need to attract their attention and introduce yourself.

Excuse me, do you have a minute? Would you mind answering a few questions?

Excuse me, do you think I could ask you a couple of questions about your ...? Thanks. My name is ... and I work at ... What I'd like to ask you is: ...

Other questions you might like to ask are:

Could you give me some more details about ...?

Where can I get more information about ...?

Can I just pick you up on something you said in your presentation?

If you want to talk to someone who has just done a presentation and you are in a line with other people, the presenter will probably want to deal with each person in the line as quickly as possible.

So, when you finally get to talk to the presenter say:

I don't want to take up your time now. But would it be possible for us to meet later this evening? I am in the same line of business as you, and I have a proposal that I think you might be interested in.

4.6 How can I introduce myself to a group of people?

To avoid having to introduce yourself into a group, you could try to arrive early at any social events. This means when you see your key person entering the room, you can go up to them immediately before they get immersed in a conversation.

If your key person (i.e. the person who you wish to meet) is already chatting to another person or a group of people, then you need to observe their body language and how they are facing each other. If they are in a closed circle, quite close to each other and looking directly into each other's faces, it is probably best to choose another moment. However, if they are not too close, and there is space between them, then you can join them. In such cases you can say:

Do you mind if I join you?

I don't really know anyone else here. Do you mind if I join you?

Is it OK if I listen in? [to listen in means to listen without actively participating]

Sorry, I was listening from a distance and what you are saying sounds really interesting.

Then you can wait for a lull (pause) in the conversation and introduce yourself:

Hi, I'm Carlos from ABC.

At this point you have their attention. You can either continue by asking a question to check that you have correctly identified your key person.

Are you Miroslav Garbarek? Because I have been really wanting to meet you.

If there is no key person in the group, but in any case the conversation seems interesting. You can say:

What you were saying about x is really interesting because ...

So where do you two work?

Thus you can either immediately start talking about what you do, or ask the other people a question. Asking a question is the most polite strategy as it shows your are interested in them. It also gives you a chance to tune into their voice.

At some point, someone in the group will probably ask you what you do. Rather than stating your position it is generally best to say something more descriptive and specific:

I work for Ferrari in R&D. I am investigating new ways to produce fuel efficient cars.

I just started a new job at ABC, where I am developing some software to enable ...

4.6 How can I introduce myself to a group of people? (cont.)

If you are more descriptive, people are more likely to make comments or ask questions. If you just say *I work for ABC,* then the conversation may then be directed to someone else.

In any case, make sure you do not spend too much time talking about yourself. Find out what the other members of the group are interested in and focus on that.

If you no longer wish to keep talking to the group you can say:

Well, it's been really interesting talking to you. I'll see you around.

I've really enjoyed talking to you. Hope to see you at our stand.

The use of the present perfect (*it has been, I have enjoyed*) immediately alerts the rest of the group that you are about to leave.

4.7   What can I say if I see a useful potential contact at the coffee machine?

If your key person is alone by a coffee machine this is a great opportunity as you will hopefully get their undivided attention.

First you need to attract the key person's attention. Here are some phrases you could use:

Excuse me. I heard you speak in the round table / I saw your presentation this morning.

Hi, do you have a couple of minutes for some questions?

Excuse me, could I just have a word with you? I am from ...

Second, it is generally a good idea to say something positive about the person and / or their work:

I really enjoyed your presentation this morning.

I thought what you said at the round table discussion was really useful.

Third, suggest you move to somewhere where you can sit.

Thank you, shall we go and sit in the bar?

Shall we go and sit over there where it is a bit quieter?

If you see that they are in a hurry, then it is best to arrange to meet later. Show that you understand that the person is busy and that you don't want to take up much of their time. In fact, tell them the exact amount of time involved, this is more likely to get them to accept.

Would after lunch suit you?

Shall we meet in bar?

When do you think you might be free? When would suit you?

Would tonight after the last session be any good for you?

Could you manage 8.45 tomorrow?

I promise I won't take any more than 10 minutes of your time.

If they agree to your proposal, then you can say:

That would be great / perfect.

That's very kind of you.

5 MANAGING AND PARTICIPATING IN A FACE TO FACE MEETING

5.1 Exploiting the few minutes before the meeting begins

Use the time while other participants are arriving to practice your English and settle your nerves (see 3.4).

5.2 Announcing the start of the meeting

If you are the chairperson, typical ways to signal that you want the meeting to start is to say one of the following in a loud voice:

OK, I think everyone is here.

Right, shall we get things moving?

Let's begin / get going / get started, shall we?

Perhaps we'd better get started / get down to business.

If you are not the chairperson and know that someone else is due to arrive you can say:

Vladimir should be here in a few minutes.

My manager has just texted me to say she will be here at 10.15. She apologizes for the delay.

I thought David was supposed to be coming.

A. Wallwork, *Meetings, Negotiations, and Socializing*,
Guides to Professional English, DOI 10.1007/978-1-4939-0632-1_5,
© Springer Science+Business Media New York 2014

5.3   Introducing the attendees

When not all the attendees know each other, it is quite common for each person to briefly introduce themselves in response to a prompt from the chairperson:

> Would you like to say a few words about yourselves?

You may wish to say: 1) your name, 2) your position, 3) why you are attending

> Hello, I am Bui Than Liem.
>
> I am the sales manager at Viet Merchandising
>
> I am here in order to …

If you are the chairperson you might decide to introduce a new person to a group of people who already know each other:

> First of all I'd like to introduce you to Ugo who is going to be working in our group. Ugo, would you like to say a few words about yourself?

5.4 Referring to the agenda, outlining objectives, talking about breaks

Although participants should already know why they are at the meeting, they may need reminding:

I've called this meeting first to ... secondly to ...

The main objective of our meeting is ...

If you are the chairperson and you have not distributed the agenda prior to the meeting, you can hand out the agenda and say:

I've prepared an outline / a rough agenda.

As you will / can see, there are five issues I'd like to discuss.

Could you look through it please.

Could you add any points you'd like to discuss.

Also, feel free to suggest any items that you think we do not need to discuss.

Alternatively, if the agenda was distributed in advance:

Have you all got a copy of the agenda?

Now let's look at the agenda in detail.

Do you have any comments you'd like to make on it?

If you are a participant, in response to the chairperson's request for comments, you can say:

No, everything seems fine.

Well actually, I was wondering why we need to discuss ...

Yes, I would like to suggest that we also discuss ...

If we have time, could we also go through ...

5.5 Announcing the time schedule and breaks

For potentially long meetings, it helps to inform participants of any breaks:

We have a lot to cover, so I suggest we have a break at 10.30 and then for lunch at around 12.30? Does that sound alright?

I have planned a break at 11.00 and arranged a buffet lunch for one o'clock. Then if it is OK with you, we can be back at the table for two o'clock.

I think we should aim to finish by four thirty at the latest.

You can also suggest the time that each item should take:

I think we can allocate 15 min to each of the first two items. Then two or three minutes to the other items.

We should take about 30 min for the first point, and around ten minutes for each of the others.

When the time comes for a break you can say:

Well, it's already 11.00, time for a break.

Perhaps before we move on to the next item we should take a break.

5.6 Opening the discussion

If you are the chairperson, you can suggest who should open the discussion:

Would you like to open the discussion......?

Perhaps you'd like to explain/tell us/give us....?

Alternatively, you can say:

OK, let's begin with item 1. Does anyone have any ideas on ...?

5.7 Bringing other people into the discussion

Once the discussion is going, you can suggest that other people join in:

Could I just bring Melanie in here, she's made a study of ...

Kaspar, would you like to tell us about ...

Sergei, I think you have been investigating ...

Some research has shown that silent participants often have a lot to contribute, as they may be silent simply because they disagree with what is being decided. To encourage such people to participate:

We haven't heard from you yet, Pierre. I would really appreciate hearing your views.

Eriko, would you like to add anything?

5.8 Moving on to the next item in the agenda

If you are the chairperson and feel that an item in the agenda has been sufficiently covered, you can say:

I think that covers the second item. We can move on to the next item.

Let's move on to the second point now.

Shall we continue then?

If you are not the chairperson, but feel that a topic has been exhausted:

Would it be alright if we move on to the next point and then come back to this later?

I think we're losing sight of what we are trying to do so can we move on to ...?

I think it might be a good idea to move on to the next point.

5.9 Interrupting and handling interruptions

Due to your English level, you may not feel sufficiently confident to interrupt someone else while they are speaking. The secret is to use a combination of body language (moving forward in your chair and perhaps raising your hand slightly) and to say 'sorry' followed by one of the following:

Could I just say something / interrupt?

Do you mind if I just say something?

I'd just like to ask Luigi a question.

If someone interrupts you and you are OK about this you can say:

Please go ahead.

That's fine, I've said everything I wanted to.

If you wish to continue speaking:

Sorry, if I could just finish what I'm saying …

Can I just finish what I was saying? It will only take me a minute.

If there is an external interruption (e.g. the phone goes, someone comes into the room, there is a loud noise), to return to what you were saying:

Going back to what I was saying / I said before …

OK, where was I? / What was I going to say?

OK, what we were saying? Oh, yes, I was saying that …

5.10 Eliciting opinions from silent attendees

A key role of the chairperson is to encourage all attendees to express their opinion so that all points of view can be taken into account and a consensus achieved. If you simply say:

Do you all agree on that?

Does anyone have any comments?

What's the general view about that?

you risk that the quieter attendees (or those whose English is poorer than the other attendees) will say nothing even if they might have something useful to contribute. Instead, it is best to ask them by name:

Katsumi, would you like to comment here?

But even the above question may not be effective because Katsumi could simply answer 'no'. So it is best to ask more direct questions.

Katsumi, what do you think would be the advantages of ...?

Shigeko, how would your department react if we took this decision?

By asking questions that require a specific answer, you ensure that attendees become much more involved in the decision-making process.

5.11 Expressing opinions

You can express your opinion or make suggestions in two main ways, by:

1) focusing on your own personal viewpoint (using *I* , *me, my)*

It seems to me that ...

As I see it .....

My inclination would be to ...

2) making it sound like a joint opinion (using *we* or no pronoun). This is a more diplomatic approach and leaves the decision more open.

From a financial point of view, it would make more sense if we ...

Why don't we ..?

What about ...?

It might be a good idea to ...

Even if you use *I*, you can still make the opinion sound less strong and more tentative:

I wonder if we could ...

I (would) recommend/suggest that we should ...

When a meeting involves people you have not met before, it generally pays to adopt a soft approach. So if you disagree with someone, it is best to avoid direct statements such as *I completely disagree* or *I can't accept that*. Instead, you can use more indirect expressions:

I'm sorry, but I have reservations about ...

Actually, I'm not sure that that is necessarily the best approach.

Also it helps if you show that you have listened to what they have said and appreciate its importance from their point of view.

I appreciate your point of view but ...

I accept the need for x, however ...

I can see why you would wish to do this, nevertheless ...

I totally understand what you're saying but ...

5.12   Making mini summaries

Whatever your level of English, it is worth checking both for yourself and on behalf of your colleagues, that you have understood what has been discussed and agreed so far. So, after each key point has been discussed, you can say:

Can I just check that I have understood what has been decided?

Could someone just summarize for me what has been agreed so far?

5.13   Taking votes in formal meetings

If you are the chairperson of a formal meeting in which a vote is required, typical phrases that are used include:

Can we take a vote on that proposal?

All those in favor. OK. All those against. Right, thank you.

So that motion has been accepted / rejected by five votes to two.

5.14 Summarizing and winding up the meeting

Typical phrases that a chairperson will say in order to wind up (conclude) a meeting are:

In conclusion ...

To sum up ...

So, if you'd like me to summarise what we've ...

So just to summarize what we've been saying ...

If the meeting is informal, and no one has offered to make a summary, you can consider offering to make a summary yourself in order to check that you have understood everything. This will avoid having to clarify misunderstandings at a later date.

Can I just summarize what we have decided, to check that I have understood everything correctly.

So if I have understood correctly, we have decided to ...

Other more informal phrases that you might hear at the conclusion of a meeting and which indicate that the speaker thinks that the meeting can be terminated, are:

I think we've covered everything so let's finish here.

I think we can stop here.

Shall we call it a day?

Shall we wind things up?

5.15   Informing attendees of the next steps

The meeting process does not end with the termination of the physical meeting. There are always some follow up tasks. The chairperson will usually indicate what he / she plans to do next with regard to:

- when the minutes of the meeting will be ready and how they will be circulated
- when and where the next scheduled meeting will take place
- what tasks, if any, attendees are expected to carry out

If the chairperson does not inform you of such details, you can ask:

(When) will you be emailing us the minutes of the meeting?

Are there any more meetings scheduled for this project?

Is there anything I / we should be doing to implement the decisions made at this meeting?

Do you want us to / Would you like us to prepare anything for ...?

5.16   Thanking and saying goodbye

When the meeting has been declared closed, typically the chairperson will thank everyone for their participation and make some enthusiastic comment about the outcome of the meeting:

Well, thank you all for coming. I am sorry we went over time, but I am sure you will agree that we have achieved a lot today. I hope you all have a safe trip home, and I look forward to seeing you again in the near future.

6 MEETINGS VIA CONFERENCE CALL AND VIDEO CALL

6.1 Preparing for the call

If you are a participant, you will vastly improve the success of the conference call if you prepare for it in advance in the same way as you would for any meeting (face to face or remote) – see Chapter 2 to learn how.

If you are responsible for the conference call:

- plan ahead so as to allow participants time to prepare what they want to see
- email information in advance – agenda, papers, background info
- emphasize the importance of participants calling in on time

The email you send announcing the call could be based on something similar to the one below.

> We will resume our weekly conference call on Mondays commencing on November 30 at 10:00 EST. The call information will be distributed early next week.
>
> The purpose of the first meeting is to …
>
> Please let me know if you have any issues or concerns.

A. Wallwork, *Meetings, Negotiations, and Socializing,*
Guides to Professional English, DOI 10.1007/978-1-4939-0632-1_6,
© Springer Science+Business Media New York 2014

6.2 Knowing how the call functions and the difficulties involved

Some conferences calls are arranged so that each participant is telephoned by an operator or secretary. When you are called, the operator will ask you to 'hold the line', which means you wait until all the participants are online.

As people join the call there is generally a beep sound. As you join, introduce yourself:

> Hi, this is Praveen. Who's on the call?

> Hi Praveen, this is Karthik. We are just waiting for Olga, Milos, Yohannes and Pei Lin.

In audio conference calls you cannot see the other participants. It will help you if you can at least 'picture' them. So if you have not met them before, try to find photographs of them: these will also indicate whether the person is male or female (you may not be able to understand this from their name).

When you are speaking, you still have to try to convey all the information that would be in your body language if this was a face to face meeting. You can do this through a combination of words and tone of voice. For example, instead of nodding in agreement you can say *I see, yes, OK, right*, and instead of beckoning someone to speak using your hands, you can say "Milos, I think you have something you might like to add".

6.3   Being a moderator

The moderator's role is to chair and facilitate the success of the meeting.

When the call starts your first job is to check who is present by checking names.

If people don't know each other, take time for a brief introduction – it can be hard talking to faceless strangers.

It is generally best not to wait for latercomers. Move on, and when they do call in, make them wait for a gap in the conversation before recapping for them.

Run through any ground rules, for example it may be helpful if people always identify themselves by name before speaking.

Ensure that everyone is clear what the purpose of the meeting is and what all the items on the agenda are.

Try to avoid throwing questions or discussions out to the group as a whole – always direct them at individuals in turn. Otherwise, people may all talk at once, and make it impossible to hear properly.

Check periodically that people haven't got lost by directing a question or comment to them.

Take notes (or allocate someone to do so) immediately from the start of the meeting, you can then use these notes to write a summary to email to participants after the meeting.

6.4   Checking the sound quality

If you are the moderator (italics in the dialog below) it is a good idea to check that no one is having any technical difficulties.

Is everyone picking up all right?

This is Milos. I can hear you fine.

This is Olga. I can't hear what you're saying – there's a high-pitched noise going on.

Is that any better?

That's fine now.

Are you on speaker phone Karthik, because everything is echoing.

Yes, I am. I'll try turning it off. Is that any better?

6.5   Establishing ground rules

It is essential to set some rules at the beginning of the call. Unlike a face-to-face discussion with multiple participants, in a conference call people cannot use body language to indicate that they wish to interrupt. This means that rules need to be made with regard to turn taking and also to avoid several people speaking at the same time. For example, if you are the moderator you could say:

> OK everyone is here now. First could I ask you all to introduce yourselves? Just your name and department will be enough. *everyone introduces themselves* We have a couple of people on the call who are not native speakers. If this call is to be successful, we need the native speakers to speak as clearly as possible. If anyone isn't sure about something please feel free to request for the information to be repeated or clarified. Also, can I just remind you all to say your name when you speak. At least the first few times. And if you ask a question, try and direct it to someone in particular.

If the moderator makes no reference to the difficulties of the non-native speakers, then it is a good idea for you to mention it immediately. For example you could say:

> Speaking on behalf of the non native speakers, I would really appreciate it if you could all speak very slowly and clearly.

6.6   Using chat facilities

Particular when there is a mix of native and non-native English speakers, using the chat facility to send messages can help resolve many difficulties. If you need to receive or make a clarification via chat, you can say:

> Sorry, I am not too clear about what Praveen said. Could you write it down for me?

> Can we just stop a second, while I write down the names of the products for you?

> Would it be OK to pause for a second and just use the chat? I think it might help us to clarify things.

6.7 Tuning in

Understanding someone on the telephone can be hard, particularly if it is the first time that you have heard that person's voice. So it is useful to dedicate a minute or two to small talk, so that everyone can get used to the sound of each other's voices. Banal questions can be used:

So Praveen, what's the weather like in Bangalore?

Olga, how did the conference go?

Here it is pouring with rain, what's it like with you?

Milos, what time is it with you?

Karthik, how was your holiday?

Yohannes, how are things going in Ethiopia?

6.8 Reminding participants about the agenda and ensuring they have all the documents required

Conference calls are often arranged at quite short notice. It is always a good idea to announce the goal of the meeting and the agenda. So the moderator could say:

The goal of this call is to discuss ...

Well, I think you know the agenda. First Yohannes is going to tell us about how much funding we can expect. Then Olga is going to talk about where we are with the draft of the proposal. And finally Praveen will update us on ...

Conference calls often involve looking at documents, so the moderator should check that everyone received them and has them to hand.

Did you all get the files I sent you last night?

Do you all have a copy of the agenda?

Have you all got the presentation open at slide 1?

Do you all have the document in front of you?

If you are not sure what is being referred to you can say:

Sorry what presentation are you talking about?

Sorry, but I am not sure I received the document.

6.9   Beginning the meeting

If you are the moderator, ensure that you make a clear verbal signal to show that you wish the meeting to start. And clearly say who you wish to begin talking.

OK, let's begin.

Right. If you are all ready I'll begin.

OK. Yohannes do you want to start?

If you are called on to start, and you think that someone else should start, then you can say:

No you go first Olga.

No, I think Karthik should probably start.

6.10   Dealing with latecomers

Below is an example of what procedure could be followed if someone joins the call when the main discussion has already begun.

Hi, Pei Lin here. Sorry I am late.

Hi Pei Lin, could you wait a second. Then I'll recap everything for you. *Moderator finishes conversation with the others* OK, Pei Lin, just to summarize what we have discussed so far. *Moderator makes a summary* I think that's everything, do any of you have anything to add?

Thanks. Sorry about being late, but for some reason I couldn't get a connection. Just before we continue, could the others just introduce themselves so that I can recognize their voices.

6.11 Ensuring you are clear when you are taking participants through a presentation or document

If you are in a teleconference rather than a video conference, it can be quite difficult for people to follow your explanations of presentations and documents. Thus you need to clearly state what slide you are on and what part of the slide you are talking about, or what page / section / line of a document you are referring to.

So, I am going to move on to the next slide now, which is slide 12.

So, we are on slide 12 now. I'd like you to focus on the figure at the top left. The one that says 'functionality'.

Can we just go back to the top of page 20.

OK, so is everyone on page 40? The middle of the page where it says 'How to set up version 2'.

6.12 Admitting that you are having difficulty following the conversation

Given the fact that there are multiple participants, it is easy to get 'lost'. This may or may not have anything to do with the fact the call is being held in English. In any case, it is always a good idea to interrupt.

Sorry, I am not sure who is talking. Can I suggest that everyone announces who they are before they speak?

This is Olga again. I'm sorry but it's hard to understand two people talking at once.

Sorry, but the line isn't great at my end, could you all speak about more slowly?

Sorry, what slide are we up to?

Sorry, what page are we on now?

Sorry, I am not sure which figure you are talking about.

6.13 Concluding the call

Typical things that people say to signal that the conference call is over include:

> I think we've covered everything, so let's finish here.

> Right, I think that just about finishes it.

> This is a good point to end the meeting.

> Yohannes mentioned that he needs to stop at 11.45, so I think we should conclude here.

> OK, I've said all I want to say, so unless any of you have anything to add, I think we can stop here.

> Has anyone else got anything they want to add?

It is a good policy to tell people what you are going to do next, and what you expect them to do next.

> I have taken a few notes, and I will email everyone with a summary of what we have decided. If I miss anything out, then please let me know.

> I'll get the minutes of this sent out to you. Praveen, if you could send in the draft proposal for funds that would be great.

Final remarks:

> Thanks everyone for making this call, particularly you Karthik, it must be in the middle of the night for you!

> Thanks for your time everyone.

> Bye everyone.

> See you next week.

6.14   Videoconferences

Most of the above subsections are also valid for videoconferences. The main difference is obviously that you can see the other participants, so there is no need to introduce yourself every time you speak. Also, if the video quality is good you can make eye contact and you can see the reactions of the other participants.

So the only extra phrases you might need are:

Can everyone see OK?

How is the sound quality for you guys?

Can you see the slides OK? Do you want me to make them bigger?

Note that if the video quality is poor it may seem that your remote interlocutor is avoiding eye contact with you, but obviously this may not be the case.

6.15 Skype calls

Skype can be used for videoconferencing. Bear in mind that:

- whether participants will be able to use the video option may depend on the number of participants and how good the internet cable is

- sound quality may vary considerably from one participant to another – again this may depend on the line, but also on their PC, and how they are positioned in front of their PC

Given the above two points, it generally makes sense to have a sound / video quality control check before you begin the actual meeting. Typical phrases you may need are:

> Vishna, your voice isn't very loud, could you turn the volume up or sit nearer the microphone.

> Neervena, I can't see you very clearly – can you see me?

> I think we might be able to improve the sound quality if we turn the video off.

> OK, given that we have the video off, could I ask each of you to announce who you are before you say something [this will only apply when there are many participants who don't know each other].

A major advantage of Skype is that you can send written messages to each other while you are speaking. So you can exploit this option if you are having difficulty understanding someone's English or when you need to clarify something that you are saying. You can say:

> Sorry, I am having some trouble understanding. Do you think you could just type the name of the product / website / document?

> Sorry, I am having difficulty saying the word. I am just going to type it for you.

For more on audio and conference calls, see Chapters 11 and 12 in the companion volume: *Presentations, Demos and Training Sessions.*

7 PREPARING FOR A NEGOTIATION

7.1 What skills does negotiating involve?

We negotiate every day of our lives: with bosses and colleagues, our family, shopkeepers etc. Similar skills needed to deal with a difficult teenager can be applied to a business negotiation. Essentially, you need to

- have a clear goal
- know what parts of that goal you are prepared to compromise on
- be flexible
- aim not just to ensure your own satisfaction but also the other party's satisfaction

7.2 How can you improve the chances of success of a negotiation even before it starts?

The more you know about your opposite party, their psychology, their culture and their priorities, the greater the chances of you obtaining what you want from the negotiation.

Cultural and psychological aspects are beyond the scope of this book, but you can find out a little about their priorities by asking some preliminary questions via email or telephone. For example:

In preparation for next month's meeting about ... I was wondering if I could ask you the following questions:

Obviously it may also pay to reveal your own position:

I just wanted to outline our main conditions for the ...

I thought it might be useful for you to know in advance our ...

A. Wallwork, *Meetings, Negotiations, and Socializing,*
Guides to Professional English, DOI 10.1007/978-1-4939-0632-1_7,
© Springer Science+Business Media New York 2014

7.3 How should I conduct a pre-negotiation over the phone?

Before any formal negotiations are conducted, often preliminary phone calls are made. The example dialog below is between someone from the sales department of an IT company and a client (in *italics*) who is interested in an agreement to license a software product (SIMDEX) of the IT company.

Good morning this is Pandit Chikra from ABC. I'm calling to see if it's possible to have SIMDEX on a trial basis.

Yes of course we could arrange that for you. You can have a one-month trial period.

Would we get the complete product or just a demo version?

It's the complete product. After the month's trial the product automatically stops working.

Is it possible to have additional features, I mean functions that are made to measure for our company?

Well you would obviously have to specify your requirements and then we would see whether it's possible to make the modifications or additions that you requested.

What if we find some bugs?

If you discover any bugs then a new version will be sent.

If after the month's trial period we decide that we would like to purchase the product, what is the procedure?

You simply send us a fax informing us that you agree to our offer. In any case at the beginning of the trial period we will send you the license agreement and the conditions of sale.

How much is the basic cost?

The basic configuration is ten thousand US dollars.

And is that a one-off license?

No, that's the annual license fee.

And is there a discount if we decide to buy several copies?

Yes, we can give you a discount, but it will obviously depend on the numbers involved.

Would you be willing to give me some names of companies who are already using your products?

Well, I'd have to check with them first, but yes it should be possible.

Going back to the cost. Does the ten thousand dollars include subsequent upgrades or are they charged separately?

7.3 How should I conduct a pre-negotiation over the phone? (cont.)

No you will not be charged for any upgrades.

OK. Well I think that's about everything.

Well if you have any further questions then just give me a ring.

OK. Thanks very much for your help.

You're welcome. Goodbye.

Bye.

The dialog above highlights the following points with regard to the potential client (i.e. the person interested in purchasing a product or service). If you are the client, then it is a good idea to:

- prepare a list of the questions you want to ask. Remember that asking questions generally requires the use of auxiliary verbs or an inversion of subject / object. This may be difficult to remember, however if you write the questions in advance of the phone call, you will have more time to formulate them correctly.

- prepare a list of useful phrases that you might need e.g. *going back to* (to refer back to something that was discussed before), *I think that's about everything* to indicate that you have concluded your questions.

7.4 Is it worth simulating important telephone calls and negotiations in preparation for the real thing?

If you are a service / product provider, then it is useful to simulate and record (i.e. with audio) conversations such as the one in the previous section with a colleague or with your English teacher. You can do this either in your own language or in English. You can then transcribe / translate what you said, and make improvements to it. Possible improvements include:

- giving more precise and concise explanations
- giving more details in case these are required
- perfecting the English grammar, syntax and vocabulary
- choosing words that are easy for you to say

You should try and do this with several people so that you can collect all the possible questions you might be asked.

if you have prepared the questions in advance, you:

- are more likely to hear them and understand them when they are asked
- will seem very professional because you will have clear concise and detailed answers to the questions
- will be fluent and confident when you speak

8 MANAGING A NEGOTIATION

8.1 Stating your position and outlining your goal

You will normally begin a negotiation by stating your position, which could for instance be expressing your interest in buying or selling some service or product:

Our position is as follows:

We would like to purchase ...

We would be interested in selling ...

Alternatively, the aim of the negotiation might be to reach a decision or agreement.

The aim of this negotiation is to solve the problem over ...

We would like / need to reach agreement about ...

We are keen to make a decision about ...

Alternative words to *aim* are: *goal, objective, purpose,* and *target.*

A. Wallwork, *Meetings, Negotiations, and Socializing,*
Guides to Professional English, DOI 10.1007/978-1-4939-0632-1_8,
© Springer Science+Business Media New York 2014

8.2 Clarifying expectations and interpretations

When two parties sit down to negotiate, they may think that they have a clear idea of what the other party wants. However, it pays to clarify your own position and also to check your understanding of the other party's position.

Our understanding is that you are interested in ...

So, are we right in thinking that you ...?

Can I just verify that you are ...

Can we just check that we both have the same interpretation of ...

Sometimes, you may understand the other party's general position, but not the exact details of what they have in mind.

We understand you position on ...

... but what would you expect from us in terms of ...?

... but precisely what figure are you thinking of?

We read in the document you sent us that ...

We have heard from one of your customers that ...

... so are your normal prices subject to...?

... so what exactly are your normal delivery terms ...?

8.3 Making a convincing case for your product or service

When you are the provider of a product or service, you often want to make what you are offering sound like a very good deal for the customer. You can do this by using

1) words such as: *in addition*, *not only*, and *what's more*.

In addition to the 15 % discount, we are *also* willing to ...

Not only will we give you a discount, but *also*... [Note the inversion after *not only*]

2) adjectives that suggest great benefit to the customer.

We are able to quote you very advantageous terms.

We are sure you will agree that this an exceptional offer.

3) words and expressions that make it sound like the customer is to get a very special deal.

This is an exclusive offer for you.

This is the first time we have ever made such an offer.

We will make an exception in your case.

8.4 Giving concessions: use of *although, however, despite* etc

In a negotiation you often want to counterbalance something that you have said that is potentially negative-sounding with something positive-sounding, or vice versa. To do this you will need adverbs such as the following:

> *However,* you expect us to provide transport and insurance.

> *Nevertheless,* we can offer an initial discount of 5%.

> *But* we are prepared to reduce the total price by 5%.

When giving a concession, you generally stress

- the auxiliary verb (*is, are, have, could, may* etc)
- the adverb of concession (*however, although, but* etc)
- words that are used to make a contrast (*this / that, next / previous, before / after* etc)

The above words are highlighted in italics in the examples below.

> *Although* we cannot offer you a discount, we *are* prepared to extend our warranty service by six months. (1)

> *Unfortunately* we cannot carry out on-site maintenance of our software, however we *do* have a free help desk service. (2)

> OK, we agree not to have a discount, but we *do* insist on receiving delivery within five working days of placing the order. (3)

> Although at *this* point we are only interested in giving you a relatively small order, if we are happy with the products we will be placing a larger order. (4)

The examples above highlight that:

- *do* is used in an affirmative form to give extra strength to the verb it is next to (examples 2 and 3)
- in the second part of the phrase if there is an adverb of concession, then the auxiliary verb following it will generally have the most stress (i.e. you will say this word louder and longer than the other words). The adverb of concession may in any case be stressed by pausing on it for longer than the other words.

8.5   Trying to get a better deal

There are several tactics you can use to attempt to get a better deal.

1) raising potential problems:

> You've promised a tight delivery schedule but what happens if you fail to deliver on time?

> If there are any defects what compensation will you give us?

2) mentioning lack of clarity or certainty:

> We are still not convinced by ...

> How can you be sure that ...?

3) showing your appreciation but then qualifying it:

> We are pleased that you can ... however, ...

> We appreciate your efforts to ... nevertheless ...

4) refusing to accept what is on offer by saying what the negative consequences will be:

> If we accept these prices, then we will have to pass on the cost to our customers thereby risking losing a significant number of orders.

> It would not be good for our business, if we accepted the terms you currently have on the table.

5) indicating that you will not change your position

> I am afraid that we cannot change our offer.

> So, this is our final offer.

To learn how to use the conditional forms see 9.11.

8.6 Concluding the negotiation

Negotiations tend to be concluded in the following phases:

1. agreeing to and accepting the terms and conditions
2. summarizing exactly what has been decided
3. expressing satisfaction
4. discussing the next step
5. saying goodbye

To learn useful phrases connected with these five steps see 15.4.

9 KEY TENSES WHEN NEGOTIATING AND WHEN DESCRIBING YOUR COMPANY

This chapter outlines how specific tenses are used in a negotiation or meeting scenario. It is not meant to be exhaustive, but does contain a few examples from social situations as well.

This chapter is organized as follows:

9.1 Present simple

9.2 Present continuous

9.3 Non use of continuous forms

9.4 *Will*

9.5 *Be going to*

9.6 Future continuous

9.7 Present perfect simplel

9.8 Present perfect continuous

9.9 Non use of present perfect continuous

9.10 Past simple

9.11 Conditional forms

9.1   Present simple

Use the present simple for:

− states and situations that don't change.

The company only *employs* graduates – this is our policy.

I'm afraid we *don't offer* discounts greater than 15%.

Our office *is closed* at weekends.

− habits and things that are done regularly.

Who *deals* with drawing up the contracts in your company?

How often *do you go* on holiday? I *go* about twice a year.

A. Wallwork, *Meetings, Negotiations, and Socializing,*
Guides to Professional English, DOI 10.1007/978-1-4939-0632-1_9,
© Springer Science+Business Media New York 2014

9.1 Present simple (cont.)

- reporting what other people have told us or what we have found in a document.

I have spoken to my boss about offering you an additional discount and she *says*.

Clause 2 of the contract specifically *stipulates* that.

The present simple is also used with certain verbs in a formal context:

I *regret* that we will not be able to sign the contract until next week.

I *appreciate* the fact this will delay the signing, but ...

I *realize* that you would prefer to operate in this manner, however ...

Do not use the present simple for:

- making suggestions, asking for advice or offering to do things. Use *shall* or *will* instead.

Shall I email you to confirm the modifications made to the contract?

Shall we adjourn the meeting until tomorrow?

Shall I open the window?

I *will let* you know what my boss says as soon as he emails me.

- actions or situations that began in the past and continue into the present. Use the present perfect instead.

I *have lived* here for six months. [NB Not: I live here for six months]

9.2   Present continuous

Use the present continuous for:

- an action that is going on now at this moment.

  So *you are saying* that you cannot agree to Clause 5, is that correct?

  I am just texting my boss to hear what she thinks about this.

- an incomplete action that is going on during this period of time or a trend.

  The number of people using our products *is growing* steadily.

  We *are working* on a new project with ABC.

- a temporary event or situation.

  I usually *work* from the office, but this month I *am working* from home.

- future programmed arrangements. In the question form, it does not matter whether or not you know if your interlocutor has made plans.

  When *are you leaving*? I *am leaving* after the meeting this afternoon.

  I *am seeing* the CEO on Monday, we have an appointment for 10.00.

9.3   Non use of continuous forms

The types of verb below are not generally used in the continuous form (i.e. present continuous, past continuous, present perfect continuous). They describe states rather than actions.

Verbs of opinion and mental state: e.g. *believe, forget, gather, imagine, know, mean, notice, recognize, remember, think* (i.e. have an opinion), *understand*

>   I *gather* you have been having some problems with the software.
>
>   *Do you agree* with what I am saying? Yes, I *agree*.
>
>   I *assure* you / I *guarantee* / I *promise* I will be on time.
>
>   I *imagine* you must have had a long journey to get here.
>
>   I *hear / understand / gather* that you wish to revise Clause 8b.

Verbs of senses and perception: *feel, hear, see, seem, look, smell, taste*

>   This fish *tastes* delicious.

Verbs that express emotions and desires: e.g. *hate, hope, like, love, prefer, regret, want, wish*

>   *Do you want* a hand with that?
>
>   I *prefer* the sea to the mountains.

Verbs of measurement: e.g. *contain, cost, hold, measure, weigh*

>   This table *contains* the data on xyz.
>
>   The recipient *holds* up to six liters.

When the above verbs refer to states rather than actions, they may be used in the continuous form. Examples.

>   We *are having* dinner with the team tonight. [*have* means 'eat' not 'possess']
>
>   We *were thinking* about contacting them for a collaboration. [*think* means 'consider' not 'have an opinion']

9.4   Will

Use *will* to:

- respond to a request.

    A: Could you have a look at the revised draft and tell me what you think of it.

    B: OK, *I'll do* it tomorrow morning.

    A: I was wondering whether you might be able to schedule a further meeting next week.

    B: OK, *I'll have* a look at my diary when I get to the office and *I'll let* you know when will be a good time for me.

- respond to a situation that presents itself at that moment.

    My mobile's ringing. *I'll just have to* answer it.

    A: I don't really understand. B: *I'll try* to explain better. *I'll give* you an example.

- convince your counterpart.

    With our software you *will be able to* reduce costs by up to 25 %.

    Our services *will help* you reach a much wider market.

- talk about future states and events with verbs that don't take the present continuous

    We *will know* tomorrow.

    She *will be* 50 next week.

- indicate formal events.

    Our company *will celebrate* its 50th anniversary next year.

    The meeting *will take place* at 10.00 in Room 6.

- make requests.

    Will you give me a hand with this translation please?

    Will you let me know how you get on?

9.5 Be going to

This form is not used frequently in negotiations or meetings, but more often in less formal situations. Use *be going to* plus the infinitive to refer to plans and intentions that:

- you have already made decisions about but for which you have not necessarily made the final arrangements:

    She's going to try and get a new position at the London office. [This is her plan but she hasn't necessarily started to look yet]

    Are you going to see Buckingham Palace while you're in London? [Is this part of your planned itinerary?]

- do not involve making arrangements with other people.

    After the meeting *I am going to* have a long bath back at the hotel.

    Tonight I *am just going to* read through my notes, then I *am going to* go to bed.

9.6 Future continuous

Use the future continuous to:

- give the idea that something will happen irrespectively of your own intentions or wishes. There is a sense of inevitability – the future continuous implies that something is beyond your control.

    I'm sorry but *I won't be attending* your presentation tomorrow [This gives the idea that the decision does not depend on you but unfortunately there are more urgent tasks that require your intention].

    I'll be going to the station myself so I can give you a lift there if you like [This gives the idea that I am not doing you a personal favor by taking you to the station, in any case I have to go there myself, it is slightly more polite than saying I am going to the station]

- talk about plans and arrangements, again when you want to give the sense that your actions do not strictly depend on you. The implication is that this is simply the way things are.

    The CEO *will be arriving* on the 10 o'clock flight.

    As of 15 January we *will be increasing* the cost of our delivery service.

- give the idea that you have already been working to make something happen.

    I *will be sending* you the report next week. [This sounds like you had already made the decision independently of the current request by your interlocutor]

    I *will send* you the report next week. [This sounds like you made the decision now as a reaction to your interlocutor's request]

9.7 Present perfect simple

The Present Perfect often connects the past to the present. The action took place in the past but is not explicitly specified because we are more interested in the result than in the action itself. Use the present perfect for actions that took place:

- during a period that has not yet finished.

    So far *I have interviewed* two out of three candidates. [I still have time to interview the third candidate]

    I've written three reminders to them. [And I will probably write another reminder too].

    I just wanted to check whether you *have received* any news from IBM. [I don't know if you have received news yet]

    Did you receive my last email message sent on 10 March? [Precise date given]

- at an indefinite or unknown time.

    I've read three books on this subject.

    I have been informed that ...

    I'm sorry I *haven't replied* earlier but I *have been* out of the office all week.

Use the present perfect also for:

- actions or states that began in the past and continue into the present

    I *have worked* here for six months. [NB Not: I work here for six months].

    We *have not made* much progress in this project so far.

- specifying what is new and to indicate what actions have been taken.

    I have looked at your report and *have just added* a few comments. Hope they help.

    I have spoken to our administration department and they *have forwarded* your request to the head of department.

- when something is happening for the first (second, third etc) time.

    This is the first time we *have encountered* such a problem.

    This is the second time I *have been* to Caracas.

9.8 Present perfect continuous

Use the present perfect continuous to:

- describe actions and trends that started in the past and continue in the present.

    We *have been expanding* into Asia for the last three years.

    I've been revising the draft all morning, but I still haven't finished.

    Interest rates *have been going up* all year [and have not stopped going up].

- talk about the effect of recent events.

    Why are you covered in ink? *I've been repairing* the photocopier.

    He's been working for 14 hours nonstop that's why he looks so tired.

- outline problems or to introduce a topic.

    I gather you *have been experiencing* problems in ordering our products.

    I've been talking to Jim about the fault in your computer but I can't find your e-mail describing ...

9.9 Non use of present perfect continuous

Do <u>not</u> use the present perfect continuous for completed actions or when you talk about the number of occasions that something has happened or when you specify a quantity [except in days, hours, minutes etc]. Use the present perfect simple or past simple instead. Compare:

We *have been writing* a lot of reports recently. [And we are likely to write some more].

We *have written* six reports in the last three months. [The next report will be the seventh, the action of writing the first six reports is over]

I *have worked* on several projects in this field. [These several projects are now finished, but I am likely to work on similar projects in the future]

I *have been working* for three years on this project. [This project is still ongoing]

I *worked* on three projects in that field, before switching to a completely new line of business. [I now work in a different area]

He's *been talking* on the phone all morning. [And he is still talking now].

I've talked to him and we've resolved the matter. [The discussion is over]

Note the difference between the simple past, the present perfect, and the present perfect continuous:

I *have been trying* to call you. [And I will probably continue calling you]

I *have tried* to call you. [Probably recently, but I've stopped trying]

I *tried* to call you. [At a specific moment e.g. this morning, yesterday, at the weekend, I will not try again]

9.10　Past simple

Use the simple past to talk about completed actions in the recent past (even one second ago) or the distant past.

Our company *was founded* in 2013.

Last year we *made* over $6,000,000 in profits

The sales manager *called* this morning to verify …

Even if the precise moment is not mentioned, but this moment will be clear to the listener, use the simple past.

Regarding the data you asked for, I *forgot* to mention that …

I have inserted that new clause that we *talked* about.

The use of the simple past or present perfect may depend on when we are talking.

We've made a lot of money this first quarter. (March – the first quarter of the year hasn't finished)

We made a lot of money this first quarter. (April or later – the first quarter is complete)

We use the present perfect to talk about actions in an indefinite past. But when we ask for or give further details about those actions we use the past simple.

Have you ever bought anything from Amazon?

What exactly did you buy? How long did it take to receive them?

We use the present perfect for past actions that have results in the present.

I've bought so many books that I don't know where to put them.

I can see that you've made a lot of progress.

9.11 Conditional forms

In negotiations two types of conditional form are typically used:

Type 1: *if* + present + *will* or *will* + *if* + present

Type 2: *if* + past + *would* or *would* + *if* + past

But you may also occasionally need Type 3:

Type 3: *if* + *had* + past participle + *would have* + past participle

or: *would have* + past participle + *if* + *had* + past participle

Instead of *will*, sometimes *going to* is used, and *would* is sometimes replaced by *could*.

Use Type 1:

− when you are making certain offers and you say what the probable result of that offer will be. It sounds like action is more likely to be carried out, so it is more persuasive than Type 2 conditionals.

If you order more than 10, we *will give* you a discount.

− to confirm facts or to offer concessions that we are genuinely willing to make.

What *will* the cost be *If we go* for the basic model?

− to talk about real future situations

If I have time *I will try* to finish the draft this evening.

Use Type 2:

− when you are making tentative offers or when we want to show we are more hesitant or less willing to make a concession.

Would you accept our offer *if we agreed* to lower the price by 1 %?

− to talk about improbable or unreal future situations, or when making cautious suggestions.

If they offered me a million dollars, I still *wouldn't accept* the job.

How soon *would you be able* to make delivery, *if we accepted* your offer today?

Use Type 3

− to expresses how things might have been if something had (not) happened. It can be used to express regrets and hypotheses about the past, missed opportunities and criticisms of oneself or others.

If I we *had known* about this before, we *could have given* you an additional discount ... but unfortunately now it is too late.

In the case of this specific client, they told us that *would never have* entered the Asian market if they *had not been able* to use our services.

10 GENERAL RULES OF SOCIALIZING

10.1 What skills do I need to have a successful conversation?

To have a successful conversation and consequently to improve your business and networking skills you need to be able to:

- break the ice i.e. initiate a conversation with someone you have never met before or who you only know superficially

- embark on safe topics

- carry forward the conversation i.e. allow the conversation to move forward in a logical and friendly manner with no long silences

- take turns in talking i.e. never dominate the conversation

- get the other person to feel important by encouraging them to talk about themselves

- listen carefully

- collaborate with their interlocutor to fill silences by referring back to something that was said earlier

- react sensitively to what is being said

- contribute and make the right comments / noises when someone is telling a story

- know when and how to end a social interaction.

Your aim is to create an interpersonal bond, which you can then exploit in your business relations.

When speaking in another language we tend to forget the social skills that we have in our own language. However, these skills are imperative for successful business and social encounters.

A. Wallwork, *Meetings, Negotiations, and Socializing,*
Guides to Professional English, DOI 10.1007/978-1-4939-0632-1_10,
© Springer Science+Business Media New York 2014

10.2　What makes a successful conversation?

Different people from different cultures have different ways of conversing. Even men and women of the same nationality converse in a different way. Various researchers in the US and UK have shown that in many countries in the West, women tend to disclose more about themselves than men, and men tend to focus more on their accomplishments and sport. Women often use more words and give more details than men, with the consequence that men 'tune out'. In a work environment women tend to take things more literally, and men tend be more lacking in sensitivity. Finally, men tend to interrupt more in a discussion or conversation than women.

The dialog below is an example of what in many countries would be considered as an unsuccessful conversation. The two speakers have never met before and they are waiting for a presentation to begin at a conference.

Sorry, is this seat taken?

No.

There are a lot of people here for this session, aren't there?

Yes, there seem to be.

Do you know the presenter? I think she is from Harvard.

Yes.

Hi, my name's Eriko Suzuki, I work for a Japanese pharmaceutical company. And you?

I'm in medical research.

What kind of medical research if I can ask?

Smoking related diseases.

Really? That's interesting because we are developing some medicine to help smokers stop smoking.

Oh.

I work in the research department there and we are looking for collaborations. *pause* So is this your first time in Istanbul?

No, I have been here many times.

Many times?

Yes.

Oh, I have just seen a colleague of mine over there. Bye.

10.2 What makes a successful conversation? (cont.)

The dialog is exaggerated, but highlights a common problem in conversations—ones that are completely one-sided. The woman (in normal script in the dialog) is trying to be friendly, but the man (in *italics*) rejects all her attempts at getting the conversation going. It may simply be that the man is shy and / or is worried about not speaking good English. But the impression he gives to the woman is that he simply does not wish to communicate. This leads to a breakdown in the communication and the result is that the man misses a possible opportunity to collaborate with the pharmaceutical company where the woman works.

Below is a different version of the same dialog. Note how the two speakers:

- immediately start a friendly conversation
- share experiences
- show interest in what the other person is saying
- repeat back the same question that they have been asked
- repeat back what their interlocutor has just said to encourage him / her to continue
- avoid dominating the conversation and take equal responsibility for its success
- interrupt a pause in the conversation by referring back to what the other said earlier.

The context of the dialog is the same, but the dynamics are very different.

Sorry, is this seat taken?

No sorry I just put my bag here that's all. I'll just move it so you can sit down.

Thanks. There are a lot of people here for this session, aren't there?

Yes, I think we are all here to hear the professor from Harvard, she's supposed to be really good.

Yeah, I have read a lot of her papers. Really excellent. Have you come far to be here?

Well not too far, from Cairo actually. And you?

From Cairo wow! I've come from Tokyo I work for a pharmaceutical company. I'm Eriko, by the way.

Ahmed. Pleased to meet you. So you work for a pharmaceutical company?

Yes, I am in the research department. We are developing an anti-smoking drug.

Well that's a coincidence. At my lab we are working on smoking-related diseases.

10.2   What makes a successful conversation? (cont.)

Well I must introduce you to my boss, he will be interested.

Great idea, maybe you could introduce him to me after this morning's sessions.

Yeah, definitely. So you were saying you are from Cairo, do you mean you were born there?

The keys to a successful conversation are:

- take equal responsibility for keeping the conversation going
- introduce new topics naturally—don't jump from one topic to another
- link what you say to what other person has just said
- show interest.

10.3 What can I talk about when I have just met someone for the first time?

The initial exchanges people have when they have just met are known as small talk. These include non-risk topics such as:

* the weather

* the town or country where you are now

* the hotel where you are staying

* how you traveled to where you are now.

Typical questions that people ask while making small talk are:

> It's a bit cloudy, isn't it?
>
> Do you think it's going to rain later on?
>
> Did you have to travel far to get here?
>
> Which hotel are you staying in?
>
> Have you been here before?

Cultures differ considerably in the way they attempt to establish a relationship using small talk. In much of Europe and North America, initial conversations often focus on the person's job, and in Japan on the organization they work for. Whereas Arabs may initially attempt to find out about each other's family identity.

Such exchanges enable you and your interlocutor to:

* get used to each other's accents and style of speaking. You are not giving each other essential information, so it does not matter at this point if you do not understand everything you say to each other

* find your voice in English

* make a connection with each other

* learn a little personal information that you might be able to refer to in future conversations

* make some positive comments about each other. This positive feeling will then be useful if any negative comments need to be made later on (for example in a technical discussion).

10.4 What are the typical safe topics that involve cultural similarities rather than differences?

Casual meetings at bars and restaurants at international trade fairs provide a perfect opportunity for discussing similarities and differences in culture. If you focus on the similarities this will generally create a better atmosphere, rather than trying to claim that your country does things better than another country.

This does not necessarily involve having heavy ethical or political discussions but can be centered on more straightforward, but nevertheless interesting, topics such as:

- legal age to do certain things (e.g. drive a car, vote)
- dialects and different languages within the same national borders
- the role of the family (e.g. treatment of the elderly, ages people leave home)
- things people do for fun (e.g. bungee jumping, karaoke)
- tipping habits (e.g. hotels, restaurants, taxi drivers)
- holiday destinations
- jobs and how often people change them, how far people commute to work
- national sports
- natural resources
- shop and office opening and closing times
- punctuality and its relative importance.

If you prepare vocabulary lists for the above topics and learn the pronunciation of the words, then you will have more confidence to initiate and / or participate in a conversation.

10.5 Are there some topics of conversation that are not acceptable for particular nationalities?

There are some topics of conversation that are universally acceptable, such as those used for breaking the ice. However, money is a topic that some British people might consider inappropriate for discussion with strangers at a social event—this means that they might find it embarrassing to be asked questions about how much they earn, how much their house is worth, how much they spend on their children's education.

What is appropriate varies from nation to nation. A Japanese woman told me:

> In Japan we are hesitant to talk about personal matters. For instance, many British people I have met like to talk about their families and show photographs, but the Japanese don't do that, at least not in depth. We would say "I have a husband. I have a son and I have two daughters". Japanese men like talking about hobbies, golf, for example. We talk about food. Women even like to talk about what blood type they are.

Sometimes you may think that your interlocutor is asking too many questions, which may be also too personal. Most Anglos would not consider questions such as *Where do you work? What did you study? What did you major in? What seminars are you planning to go to? Did you take your vacation yet?* to be too personal. Such questions are merely a friendly exploration in a search to find things that you may have in common. The purpose of the questions is merely trying to find some common ground on which to continue the conversation.

Some questions would be considered inappropriate by most Anglos, for example:

How old are you?

What is your salary?

What is your religion?

Are you married?

How old is your husband / wife?

Do you plan to get married?

Do you plan to have children?

How much do you weigh?

Have you put on weight?

How much did you pay for your car / house (etc)?

10.6 What kind of topics are generally not of interest to the interlocutor?

If you want to be a successful networker and be able to set up new collaborations, then it helps if you can talk about things that will be of interest to your interlocutor. Imagine the topic of conversation is holidays. A lot of people may not necessarily be interested to hear what hotel you stayed in, what museums you visited, how much the metro cost etc— unless of course they are planning to go there themselves. They are more interested in holiday disasters: planes rerouted or cancelled, luggage lost, food poisoning.

By listening and analyzing the conversations going on around you, you should be able to get a clearer idea of what topics people find interesting, and more specifically, what aspects of those topics generate interest.

10.7 If my company is hosting visitors, what are the typical non-work questions that guests might ask me?

If your company is hosting some foreign visitors then you have the perfect opportunity to share your knowledge of the local area, and also to practice your English! Here are some typical questions and answers:

Are there any good restaurants where I can try / sample the local food?

Yes, there is a good one near the town hall, and another one just round the corner from here on Main Street.

What local sites would you recommend that I go and see?

Well the standard places where all tourist go are .... But I suggest that you visit the museum of ... and if you like food you could go to the market on Main Street.

Do you have any suggestions as to where I might buy a ...?

You could try the department store which is on the main road that leads to the mosque.

Note the construction with suggest and recommend: to suggest / recommend that someone do something.

If someone is critical of something (e.g. poor service in your country or company), and if you don't want to enter into a long defense, you can simply say:

Yes, I know what you mean.

Or if you want to be more defensive you can say:

Well, to be honest, I just think you have been unlucky.

10.8 How can I prepare for the social conversations over business lunch / dinner?

You will probably be able to participate more effectively in a conversation if you initiate the topic area yourself. You could prepare short anecdotes on one or more of the following:

- travel stories (e.g. missing planes, terrible hotels)
- the worst presentation you ever did
- the best / worst trade fair or conference you ever attended
- new technologies.

These are good topics because they are neutral and everyone in your group is likely to have something to contribute. If you initiate the conversation, it will help to boost your confidence.

An alternative to stories / anecdotes are factoids (i.e. interesting statistics), for example factoids about your country, about your field of business, or about anything you find interesting.

It is also helpful to learn something about psychology and communication skills. Socializing is all about relating to people and communicating well with the other attendees. Learning good communication skills and social skills entails knowing how the human brain receives information, and how we perceive each other.

10.9 How do I decide how formal or informal to be?

A frequent cause of misunderstanding and embarrassment is when two people expect a different level of formality from each other.

Let's imagine a meeting between a Spaniard and a Bangladeshi, both with identical roles within their respective companies.

The Spanish guy is probably accustomed to conducting social exchanges on an informal and friendly level. He would be surprised if his interlocutor referred to him as 'Sir' during such a conversation. In such cases, he might feel that his interlocutor is putting him on a superior level, which he does not feel comfortable with.

By contrast, the Bangladeshi may perceive the Spaniard as trying too hard to be friendly and this may make him feel uneasy. The Bangladeshi is used to showing people respect and in return being shown respect.

There is no easy solution to this very common situation when different cultures meet with differing ideas about the norms of communication. Both the Spanish and the Bangladeshi are conducting a conversation following their own norms. The secret is probably just to be aware that we don't all socialize in the same way, and to try and adopt some neutral middle ground where we are neither too friendly or too formal, and where both parties feel comfortable.

10.10 How can I practise my English grammar in a social context?

Talking in a social context is an excellent opportunity to practise your English in a situation which is generally relaxed and not critical from a work / career point of view. You will have the opportunity to ask a lot of questions, and you can phrase these questions using many different tenses (see Chap. 9). Examples:

PRESENT CONTINUOUS What are you doing here in *name of town*?

PRESENT PERFECT CONTINUOUS How long have you been living in *town*?

FUTURE CONTINUOUS How long will you be staying for?

GOING TO How long are you going to stay in *town*?

PAST SIMPLE Why did you choose *town*?

PAST PERFECT Had you thought about going anywhere else instead of *town*?

THIRD CONDITIONAL If you hadn't come to *town* where would you have gone?

10.11 I am too shy and embarrassed to have a conversation in English, what can I do?

One common reason for an unsuccessful conversation between two people is that one of the people fails to contribute to the conversation because they are too embarrassed about their level of English or because they don't talk much even in their native language.

Do you like standing up in front of other people or do you feel nervous and self conscious? If you are the kind of person who usually does not talk much at dinners, parties and even in everyday banal social situations (e.g. in front of the coffee machine, on the telephone), then try and make an effort to talk more and find yourself at the centre of attention.

Other ways to become used to being the center of attention or at least to have people focus on you include:

- joining a dance or acting group
- offering to do presentations at work
- talking to strangers sitting next to you on trains and planes
- sports coaching for children
- doing voluntary work

Don't just listen to people, learn to have the courage to interrupt them and comment on what they have said. For instance you can relate what they have said to your own experience. You could say:

I know exactly what you mean. In fact …

Actually I had a very similar experience to what you have just described.

I was once in exactly the same situation.

I completely agree with what you are saying. In fact, …

I am not sure I totally agree with you. In my country, for instance, …

Tell people things that have happened to you or that you have read or heard about. You can do this in low risk situations (i.e. where your conversation skills and level of English are not going to be judged), for example, when you are with a group of friends.

You could practise doing two-minute presentations with a group of work colleagues. You could either do this in your own language or in English.

10.11 I am too shy and embarrassed to have a conversation in English, what can I do? (cont.)

Possible topics:

- what you enjoy doing most in life
- your favorite movie or book and why you like it so much
- the worst journey of your life
- the best holiday
- your dreams for the future
- your ideal house

If you practice being at the center of attention you will gain more confidence.

10.12 My English vocabulary only extends to a limited number of topics—what can I do?

Sometimes your ability to participate in and contribute to a conversation will depend on the vocabulary you have available on that particular topic. If you feel you don't have the vocabulary required you could try to gently shift the conversation to an area where you know a greater number of relevant words. Of course, this shift must be to a related area rather than a totally new topic, unless there is a complete silence where it would be justified to change topic.

Food is often a subject at lunch or dinner, regarding not only the menu of the meal itself, but also discussions about the national and typical dishes of those around the table. Discussing such dishes involves a lot of specialized vocabulary regarding ingredients and cooking techniques. However, there are other aspects of food that also have a strong cultural interest. You can inject considerable interest into a conversation about food, if you talk about the social aspects of food and eating, rather than just typical dishes. For example, you could discuss:

- taboos—i.e. what foods are not acceptable to be eaten by humans (e.g. in the UK, horsemeat is rarely eaten, and cat and dog meat are never eaten)
- fasting—i.e. what foods are prohibited for religious reasons at certain times of the year
- events—i.e. what foods people eat on particular occasions (e.g. in the USA it is common to eat turkey to celebrate Thanksgiving)
- etiquette—i.e. how guests are expected to behave (e.g. can you refuse if your host offers you more food? should you take a gift, if so what is and is not appropriate? should you take off your shoes before entering someone's house)
- production methods—i.e. genetically modified foods
- the pros and cons of being vegetarian
- food allergies

There are three ways to do this:

- Wait for a pause in the conversation and initiate a change in topic by saying: In my country at this time of year, we can't eat meat …
- Invite others to begin a discussion by saying: I am curious to know whether anyone else is allergic to …
- Ask a question: In your country do you have many vegetarians?

The result of this is that you will find social events more rewarding and less frustrating. Also, people will see you as someone who is able to manage a conversation and make useful contributions. These two skills are obviously also applicable outside a strictly social situation and in the context of work. Thus you will demonstrate that you are the kind of person that is easy and efficient to work with.

10.13 I find it more difficult to understand English than to speak it? Is it a good tactic to talk rather than listen?

Many non-native speakers are afraid or embarrassed about not being able to follow a conversation due to poor listening skills. One strategy that some use is to try to increase the amount of time they spend speaking. Clearly the more you speak, the less you need to understand other people.

If you adopt this 'talking rather than listening' strategy, continually check that your listeners are following you and are interested in what you are saying. If they are not giving you any eye contact, it probably means that either they cannot understand you or they have lost interest.

You may compromise your chances of future collaborations if you are seen to dominate conversations. The solution is to accept the fact that you will not understand everything and as a consequence let all the group of people you are with talk in equal amounts.

11 MANAGING A CONVERSATION

11.1 How should I introduce myself?

Most Anglos today introduce themselves in a very simple way by saying:

Hi, I'm Richard.

Hi, I'm Richard Jones.

Hello I'm Richard Jones.

Good morning I am Richard Jones.

Anglos say their first name (*Richard*) followed, in more formal situations, by their family name (*Smith*).

If someone asks *What is your name?* you would normally reply with both first and family name.

Anglos often give their own name rather than directly asking the interlocutor for his/her name. This may take place several minutes into the conversation, particularly if the conversation appears to be worth continuing. A typical introduction is:

By the way, my name is Joe Bloggs.

Sorry, I have not introduced myself - I'm Joe Bloggs from NASA.

I don't think we have been introduced have we? I'm ...

At this point you would be expected to reply with your name.

Pleased to meet you. I'm Stomu Yamashata.

If you didn't hear the name of the person you have just been introduced to you can say:

Sorry, I didn't catch your name.

Sorry, I didn't get your name clearly. Can you spell it for me?

Sorry, how do your pronounce your name?

A. Wallwork, *Meetings, Negotiations, and Socializing,*
Guides to Professional English, DOI 10.1007/978-1-4939-0632-1_11,
© Springer Science+Business Media New York 2014

11.1 How should I introduce myself? (cont.)

Don't be reluctant to ask for a repetition of the name, otherwise you will spend the rest of the conversation looking at their name tag! Also, we all like it when people remember and use our name, we feel important and consequently we are more responsive to people who remember it.

If you are too embarrassed to ask someone to remind you of their name, then you could offer them your card and hopefully they will then give you their card. Giving someone your card also means that you immediately have something to talk about:

Oh, I see you are from Tokyo, I was there last year.

So you work for ABC, do you know John Smith in the sales department?

So you work in Peru, but I think you are from Korea, is that right?

11.2 Should I address my interlocutor with his / her title?

In English there is only one form of *you*, i.e. there is no additional form of *you* that denotes respect.

If you wish to show someone respect then you can use a title. For men you can use *Mr* (pronounced *mister*) and for women *Ms* (pronounced *muz*, like the *cause* in *be*cause). *Mr* and *Ms* do not indicate whether the person is married or not.

The terms *Mrs* (pronounced 'misses') and *Miss* are not so commonly used nowadays as they indicate that the woman is married and unmarried, respectively - such information is not considered necessary for the interlocutor.

Your country may have many titles, for example, lawyer and engineer. Such titles are impossible to translate into English. This means that if you are for example an engineer, you should not address another engineer as Engineer Smith, but simply as Mr / Ms Smith or, if the person is in academia, Doctor Smith or Professor Smith. However, in emails you might wish to address an engineer whose native language is not English using the word engineer in their language, for example Herr Diplom Ingenieur Weber (for a German).

Many Anglos consider titles as being quite formal. If you use a title, they might simply say:

Please call me John.

This means that from that moment on the communication can take place in a more friendly atmosphere.

11.3 How do I move on from small talk?

Have a typical conversation often begins with the weather then moves on to other things.

It's really hot today isn't it?

Yes it is. It must be at least 30 degrees?

Is that normal for this time of year?

No, it's normally much cooler. What about in your country?

I think it's raining there at the moment?

Sorry, where are you from exactly?

I'm from ...

Really, I have never been there but I would love to go.

11.4 How can I show interest in the person I am talking to?

Everyone likes it when people show genuine interest in them - it gives them a feeling of importance and recognition. You can show interest in other people by asking questions and by showing that you are 100 % focused on listening to the answers. If you find a topic that seems to interest them more than other topics, then try to ask more questions about this particular topic. In any case, focus on questions that you think that your interlocutor will take pleasure in answering.

If you are not naturally curious about other people, a good way to think of questions is to use *how, where, why, when, what*. For example:

How did you get to the trade fair? By plane? By train?

How long are you staying for?

Where are you staying?

Where is your stand?

Why did you decide to come to this particular fair?

What are you planning to visit while you are here?

When are you going back?

When you listen to the answers, you can say *really?* which is said in the form of a question and is designed to encourage the speaker to continue. Another typical comment is *right*. For example, let's imagine that the dialog below takes place in Rome, Italy.

So where are you from?

From Stockholm in Sweden.

Oh right, so how did you get here?

By train.

Really?

Yes, I don't like traveling by plane.

Right.

And you, where are you from?

Well, I'm from Rome actually.

Oh really?

Yes, I was born here.

11.4 How can I show interest in the person I am talking to? (cont.)

Other expressions you might use are:

I see.

That's interesting.

Wow.

Fantastic.

It might feel very unnatural for you to use any of these phrases, but remember you should not say them in an exaggerated way with a lot of emphasis. Just say them in a neutral way and quite quietly. They are basically verbal noises that demonstrate to your interlocutors that you are interested in what they are saying.

11.5   Is it OK to ask very direct questions?

It obviously depends on the question. If you ask questions such as:

> What are your hobbies?
>
> What plans do you have for the future?
>
> Which football club do you support?

you are making the implicit assumption that your interlocutor has specific hobbies and specific plans, or is interested in football. Such questions are not very appropriate as conversation starters and are better rephrased as:

> What do you like doing in your spare time? Do you have any particular hobbies?
>
> So, do you have any particular plans for the future?
>
> Are you interested in football? Do you support any particular team?

However, if a topic such as football has already been introduced into the conversation then you could ask a more direct question:

> So which football club do you support?

The use of *so* at the beginning of a question helps to make the question less direct.

11.6 Can I ask personal questions relating to information I have found on the Internet about them?

Most people will be happy to talk about their work (but probably not direct questions relating to salary, bonuses etc).

If you have found out something about them from their LinkedIn or Facebook pages, then you can use that information without actually mentioning it directly. For example, if you have seen photos of a holiday in Tunisia on their Facebook page, you can say:

Have you been on holiday recently?

Are you planning any holidays this year?

But not:

Have you been to Tunisia?

How was your holiday in Tunisia?

If you mention Tunisia directly, the person will know that you have been on their Facebook page and may find this fact disturbing. But if you simply refer to holidays in general you are guaranteed an answer as they will naturally tell you about their holiday in Tunisia.

Similarly, if you notice from their photos on Facebook that they have children, you can ask *Do you have any children?* but without necessarily asking them whether they are married. The idea is not to be intrusive.

11.7 Can I offer personal information about myself as a means to ask personal questions?

If you reveal personal information about yourself, you can then find out what similarities you have with the other person. For example:

My son has just started primary school.

Oh really, how old is he? / Do you have any other children?

Yes, I also have a baby daughter, and you? Do you have any children?

11.8 What kind of questions are most effective at generating detailed answers?

Some questions could simply be answered *yes* or *no*. For instance:

So you are the assistant manager, is that right?

Did you join the company recently?

Have you been working here for a long time?

The above questions are called closed questions, because potentially the person could simply answer *yes* or *no* and thus 'close' the conversation. Typically they make use of auxiliary verbs (*did, can, are, have* etc). If you find that your interlocutor is just giving you *yes no* answers, it will soon become an effort for you to continue the conversation. So you could rephrase the questions as follows:

So I hear you are the assistant manager. What does that involve?

Did you join the company recently? What were you doing before?

So, how long have you been working for ABC? What exactly is your role?

The above are what is known as open questions, and again they make use of question words such as *how, what, which, why, when, where?*

11.9   How can I avoid jumping from topic to topic?

Avoid asking a series of unrelated questions and topics, instead link your next question to your interlocutor's previous answer. First you ask a topic question and then you ask another question (or make a comment) related to the same topic in which you ask for more details. Here is an example:

> You: So where did you go on holiday?
>
> Them: To Berlin.
>
> You: (1) *Follow-up:* So what did you think of the architecture?
>
> You: (2) *Comment:* I've heard the architecture is amazing.
>
> You: (3) *Encouragement:* So tell me all about Berlin.

Ensure your questions follow a logical order - don't just jump from topic to topic. Try to exhaust one topic before moving on to another one. There are four main ways to encourage someone to give more details on a topic that has just been initiated:

1. Restate part of what they have just said:

> Them: But the food was terrible.
>
> You: Terrible?
>
> Them: Yes, in fact we had one really bad experience when ...

2. Make mini summaries of what they've just said:

> You: So the architecture was great, but the food was terrible.
>
> Them: Yeah, and then we had a few problems at the hotel.

3. Paraphrase or agree with what they just said:

> Them: Exactly. And the meeting was so boring.
>
> You: *Agree:* Yeah, really boring.
>
> You: *Paraphrase:* Yeah, a complete waste of time. And we didn't even discuss ...

4. Show interest by asking for clarification:

> Them: And the hotel was not exactly cheap.
>
> You: What do you mean by 'not exactly cheap'?
>
> Them: Well they added on a lot of extra services.
>
> You: For example? What kind of services?

11.9   How can I avoid jumping from topic to topic? (cont.)

Them: And they had a disco every night.

You: So you're saying that it was very noisy? You didn't get much sleep. I had a similar experience last month in ...

Note how in the above exchanges the strategy is to use the clarification to initiate something that you want to say (e.g. *And we didn't even discuss...* and *I had a similar experience last month in...*). Basically you are showing respect for the other person by using a clarification to show interest in what they have said. This then allows you to take your 'speaking turn' in the conversation. If you ask for clarification this also enables your interlocutor to make any adjustments to what they said either to help you understand better or to add details.

11.10   What can I do if I find I am asking all the questions?

If you find that you are asking all the questions, then there are two possible results. One is that you may become frustrated with the attention always being focused on the interlocutor. The other is that your interlocutor might think that you are being rather invasive.

So sometimes you need to initiate a topic yourself. If for example you have been asking question's about your interlocutor's hotel, you can announce:

Well, I am staying at the Excelsior Hotel and it's not exactly cheap there either.

I know what you mean about the noise, where I am staying I get woken up at five every morning with people setting up their stalls at the market.

In the above examples you have directly related your experience to your interlocutor's experience. You also show that you have been listening carefully as you have repeated some of their concepts and phrases.

Other times you may want to initiate a completely new topic.

After work a few of us are going to a restaurant near the harbor.

Last night I went to ...

Did you hear about ...

By introducing a new topic you hope that your interlocutor will ask you some questions, and thus create a more balanced exchange. However, bear in mind that if your interlocutor seems unwilling to contribute it may have nothing to do with you - they may just be having a bad day.

11.11   What if I find that I am dominating the conversation?

There are two tactics for taking the focus off yourself.

1. Transfer their original question back to them:

Them: So, are you going anywhere interesting this summer?

You: blah blah blah. And what about you? Have you got any plans for the summer?

2. Ask them if they have had a similar experience:

You: and during the demo my laptop suddenly crashed.

Them: Oh no!

You: So I had to blah blah blah. Have you ever had any disasters like that?

11.12 How should I react to the announcements and statements that my interlocutor makes?

It is not only questions that you have to react to. People sometimes announce that they have done something. They then generally expect you to make a comment on it so that they will be encouraged to give you more details. The examples below are not connected to each other:

They: I saw a fantastic film at the cinema last night. You: Oh really what was it called?

They: My daughter has just started university. You: Oh yes, so what is she studying?

They: I have just come back from New York. You: New York. Fantastic. What were you doing there?

You also need to be able to respond to comments (rather than questions) that are directed at you.

They: You're lucky to have an airport so close to where you work.

You: Yes, it's very convenient especially with all the traveling I have to do.

They: The weather should be good when you get back home.

You: Yes, summer's in my country tend to be dry and not too hot.

They: I suggest you give my secretary a call for confirmation.

Yes: I'll do that.

11.13 Should I just reply precisely to the question I am asked, or should I provide additional information?

If you are asked a question, try to move the conversation forward by giving some extra information in your answer. For example, if you are asked the simple question: *What do you do?* or *How long have you been working here?* don't simply reply *I am a software analyst* and *3 years*.

Instead give the answer and then add more details. For example:

I am a software analyst and I work in the business to customer group. It's an interesting job because ...

I have been here for 3 years, but I was actually employed as a programmer and then last year I was promoted to analyst.

11.14   If I am in a group of people, how can I involve the others in the conversation?

There is often a tendency in a group conversation for those who speak the best English to dominate the conversation and to form a sub group. This leaves the other part of your group in silence. If your English is at a higher level than some of the others, or if you are more extrovert than them, don't use this entirely to your own advantage or as an opportunity to show off your excellent English in front of your colleagues. Instead people will appreciate it if you try and involve them. Here are some examples of how to draw people into a conversation:

Vladimir, I think you have had a similar experience haven't you?

Monique, you were telling me earlier that ...

Bogdan, I think you and Monique must be staying at the same hotel.

Domingo, I see on the program that you are doing a presentation this afternoon, what's it about exactly?

Kim, Melanie told me you are into bungee jumping.

Yoko, I read that Japan has a new government.

11.15   What do I do when there is a long silence in the conversation?

Different cultures have different tolerance levels for the length of periods of silence in a conversation. So don't think that you necessarily have to fill every silence. You can use the pause to think up new areas that you could talk about. Below are two tactics for re-initiating a conversation.

1. Return to a topic mentioned earlier or other information that you know about the person:

   So you were saying before that you had just come back from China. What exactly were you doing there?

   I seem to remember that you once lived in London, am I right?

2. Introduce new topic:

   Did you hear about that hurricane in Florida?

   So do you think Brazil will win the world cup?

   So how long do you think the boom in the market will last?

11.16 Is it rude to interrupt the other person, especially when they are doing all the talking?

Some people are used to talking a lot and having a quiet audience. For you as a listener, in a social context this may not be too much of a problem. You can simply 'switch off', look out of the window and start thinking about something more interesting. However, when having a technical discussion, informal or formal, you may wish to get your own point of view across. In such situations it is perfectly legitimate to interrupt. You can say in a friendly tone:

Sorry to interrupt you but ...

If I could just make a point ...

Just a minute, before I forget ...

Actually, I am quite curious to hear what Stanislav has to say about this.

11.17   Is it impolite to express my disagreement?

In your own language you are generally aware of when you are not being impolite. You know what little phrases you can use to sound polite. The problem of not knowing such courtesy forms in English is that you might appear abrupt or rude to your interlocutors. A native speaker may be surprised by your tone because in other contexts, for example when you are describing technical details or in writing papers/letters, you appear to have a strong command of English.

The secret is to try and show some agreement with what your interlocutor is saying before you introduce your own point of view. Let's imagine two people are discussing the relative advantages and disadvantages of nuclear power. Below are some phrases that they could use in order to express their opinions without being too forceful.

I agree with you when you say ... but nevertheless I do think that ...

You have an interesting point there, however ...

I quite understand what you're saying, but have you thought about ...

Water power definitely has an important role, but did you know that it actually pollutes more than nuclear power?

I agree with you, but I also believe that ...

The sun is certainly a safe source of energy, but ...

I know exactly what you mean, but another viewpoint / interpretation could be ...

It is not easy to be diplomatic in a foreign language, so if you do inadvertently say something that produces a bad reaction, you can say:

I am sorry, it is very difficult for me to say these things in English.

Sorry, I tend to be too direct when I speak in English.

I'm so sorry I didn't mean to sound rude.

11.18 What do I do if someone says something I don't agree with? How can I be diplomatic in my response?

If your aim is to build up a relationship in a harmonious environment then it is worth bearing the following factors in mind.

- If someone says something that you don't agree with, but the point they are making is not really important, then there is probably no benefit in contesting it

- If someone says something that is not true (but which they themselves clearly believe to be true), e.g. some erroneous data, they will probably not appreciate being confronted directly with the true facts - you will simply undermine their self-esteem

- Most people do not appreciate someone casting doubt on their opinions and beliefs, and are more likely to be even more convinced of their beliefs if these beliefs are attacked.

If you decide to disagree, then try to find some aspect of what your interlocutor has said that you can agree with. State this agreement and then mention the area where you disagree. This shows that you are at least trying to understand their point of view, and that your intentions are not hostile.

Speaker A: Your government seems to be in a complete mess at the moment.

Speaker B: I know what you mean, and there are a lot of people in my country think so too. Some progress is being made in any case. I don't know if you've heard that ...

Note how Speaker B avoids using words like *but*, *nevertheless*, and *however* (*Some progress is being made* rather than *But some progress is being made*). Frequent use of words such as *but* may put your interlocutors on the defensive and they will simply come up with more evidence to support their initial statement. This could then lead to an embarrassing argument.

If someone says something that you believe is not true, then a good tactic is to be diplomatic and say something like:

Oh really? I may be wrong, but I'd always thought that ...

I didn't know that. What I heard / read was that ...

11.19   How should I deal with questions that I do not want to answer?

If you do not wish to answer a particular question. You can say:

I really don't have any opinion on that.

That's an interesting question, but I don't think I am qualified to answer it.

That may be, I couldn't really say.

I'm afraid I don't know anything about it.

It's not really for me to say.

It depends how you look at it.

I'm sorry, I don't want to go into that.

Or alternatively you can revert the question back to the questioner:

Why, what do you think about it?

11.20   How can I check that my question will not offend or embarrass my interlocutor?

If you want to ask a question that you think might be potentially difficult or embarrassing for your interlocutor, then you can precede the question or statement by saying:

Is it OK to ask about ...?

Do you mind me mentioning ...?

Can I ask you what you think about ...?

It seems that some people in your country think that ... What do you think might be the reason for that?

11.21 How should I formulate an invitation?

Don't be too direct. For example, *do you want to go for lunch?* may make it hard for your interlocutor to make an instant decision.

Introduce the question in a more roundabout way so that you give them a chance to think of an excuse!

> I don't know what you are doing now for lunch ...

> I don't know if you have any plans for this evening but ...

> Maybe we could have lunch together because I am curious to hear about ... so you can tell me a bit about how you use ...

11.22 How should I respond to an invitation?

Below are some examples of how to respond to invitations.

Firstly, when you are happy to accept:

> I was wondering whether you'd like to sample some of the local cuisine tonight.

> That would be great, thanks.

> OK, well if I come and get your from your hotel at about seven thirty we can walk to the restaurant from there, it's just round the corner.

Another example:

> We're just going out to get a coffee at the bar, would you like to join us?

> Yes, thank you. I've just got to finish writing this email. Can I meet you there in say five minutes.

> Of course, and it's the bar opposite the entrance to this building.

Note how the person who makes the invitation also talks about the details of the arrangement.

Secondly, when you wish to refuse the invitation:

> We're having a drink at the pub on Thursday after work at about six fifteen. Do you fancy coming?

> I'm really sorry but my plane actually leaves at eight, so I really don't think I'll have the time. But thanks for asking.

11.23 What are the best ways to end a conversation that is not moving forward?

If you find that your interlocutor is failing to interact with you and that the situation is becoming awkward, then you might decide to end the conversation by making an excuse:

Sorry, I have just seen an old colleague of mine. I'll catch you later.

Sorry, I've just received an sms - do you mind if I just take a look?

Sorry, I just have to make a phone call.

Do you know where the bathroom is?

I just need to get a bottle of water. Maybe I'll see you at the presentation?

Sometimes you need to find a way out of a discussion or at least time to pause and think.

Can I just think about that a second?

Just a moment. I need to think.

Sorry, I'll have to check up on that.

Even if you have not had a long conversation, try to end on a positive note and thus leave a good impression with your interlocutor. You never know when you might see them again, or what opportunities for collaboration might arise. So, smile as you say goodbye, and say:

Well, it was nice talking to you.

Well, I hope to see you at the dinner tonight.

I'll catch up with you later.

12 ASKING WORK-RELATED QUESTIONS

12.1 How can I learn useful information about someone in a non-work situation?

When you are hosting a client it is a perfect opportunity to find out more about this person's role in their company, how they see the company progressing etc. There are often moments during a client's stay when you are not carrying out strictly work related activities. Such moments include:

- on a client's arrival at reception
- waiting for presentations to begin
- breaks during meetings
- chance meetings by the coffee machine

During such moments you can then find out information that will enable you to:

- improve your current services / products
- discover new features for your products and services that the client might like

So not only are you creating a social bond with your interlocutor but you are also extracting useful information that might be more difficult to obtain in a formal situation.

Note: In this chapter, xxx stands for the name of a product, and ABC stands for the name of a company.

A. Wallwork, *Meetings, Negotiations, and Socializing,*
Guides to Professional English, DOI 10.1007/978-1-4939-0632-1_12,
© Springer Science+Business Media New York 2014

12.2　How can I avoid seeming too direct in my quest

If you are talking in a social situation, for instance at the coffee machine, during a break or at lunch, you don't want to sound too intrusive. Instead you want to seem that you are asking questions on a casual basis. Here are some 'soft' ways to get the information you are looking for.

HARD / TOO DIRECT	SOFT / INDIRECT
How many of you are there in your group?	So I imagine there are quite a few people in your group.
Is your group getting bigger?	I know that ABC is investing a lot in e-trading so what affect has this had on the size of your group?
What is the number of senior components in your team? Which ones use xxx?	So how many senior guys are there in the group? Do you happen to know how many are using xxx?
Has anyone in your team used our product? Did they use it in their previous job?	I was just curious to know whether anyone has actually been using xxx before, you know like in their previous job?

12.3　What questions can I ask to find out about my interlocutor's current work position

You will need the simple present and present continuous to find out what your interlocutor does / is doing now or is planning for the future.

So what do you do exactly?

Are you working on anything interesting at the moment?

How long are you planning to stay with the company?

You can then move on to ask the questions that really interest you.

What plans do you have for further developments?

What's the timeline on this?

If appropriate you can reveal why you are asking such questions.

I'm asking this because I think we can help you to ...

I was just wondering about this because ...

Note the use of the continuous forms in the above two questions. They make the questions sound less direct.

12.4 How can I find out about my interlocutor's past-present work situation?

Typically, first establish how long someone has been with their current company. To do this, use the present perfect or the present perfect continuous. These two tenses make a connection between the past and the present situation.

So how long have you been with ABC?

So how long have you been working with ABC?

Use the same tenses to discover whether they have always had the same role.

So have you always worked in sales?

Have you always been doing the same job with them? Or have you been moving around?

If you establish that they are now working on a particular project, you can again use the present perfect (continuous):

How long have you been working on the project?

If you want to refer to their past experiences you will need the simple past and past continuous.

So when did you join ABC?/ When did you start work at ABC?

What were you doing before?

When were you promoted to senior manager?

For more on the use of tenses see Chap. 9.

12.5 Are there any tricks for avoiding grammar mistakes when I respond to questions?

If you are asked a question (for example those in the two previous subsections), you should reply using the same tense. Examples:

Q. So how long have you been working with ABC?

A. I have been working for ABC for 11 months / since last year.

Q. So have you always worked in sales?

A. No, I haven't always worked in sales. Before I was in marketing.

Q. How long have you been working on the project?

A. I have been working on the project for the last six months.

Alternatively you can answer the same questions but omit the verbs. This may sound more natural and also avoids possible mistake in tense usage. Examples:

Since last year.

Not always in sales. Before I was in marketing.

For the last six months.

12.6 What company-related questions could I ask?

There are many questions you can ask about someone's company. Below is just a selection.

When was the company set up?

How soon after the company was set up did you join?

When you started, did you have the same products that you have now?

Which products are the biggest sellers?

Do the biggest sellers vary much from country to country?

Which services are the most in demand?

How long did it take you to establish your brand?

What new products are being developed at the moment?

In which countries do you operate?

Are you planning to set up branches in any other countries?

How many cities in ... are you currently operating in?

What do you think makes your company different from other companies in the field?

Are their any products that are sold [any services offered] in your country that are not sold here?

Do you have to make many adaptations of your products / services for the market here?

Do you have any plans to expand across Europe?

And what about the United States, are there any plans for expansion there?

Do you think your company has a particular philosophy or culture?

Are environmental concerns important to the company?

What do you think is the most innovative thing that your company has done since you got started?

How do you see the company in ten years' time?

13 UNDERSTANDING WHY YOU DON'T UNDERSTAND NATIVE ENGLISH SPEAKERS

13.1 I get frustrated when listening to native speakers, because I try to understand everything. What is the best strategy?

Only a quarter of conversations carried out in English in the world are between native English speakers, so it might seem that understanding the English of native speakers is not particularly important. However many interesting business opportunities can be found in countries where English is the first language. Consequently, being able to understand native speakers is still important. This chapter is designed to show you why you may have difficulty in understanding native English speakers. Knowing why you can't understand may then help you to improve your listening skills.

When we are learning a foreign language we tend to think that it is important to understand everything that we hear. But when you are listening to someone talking in your own language, you probably don't listen at 100 per cent and nor do you probably need / wish to.

Thus an essential rule for improving your understanding of native English speakers is not to expect to understand everything they say.

In non-strictly technical / business encounters, conversations are often more a means of being together, a socio-cultural event in which relations are established, rather than an opportunity for exchanging information. Most of the time, what is said may be completely irrelevant. Quite often talking is merely an end in itself. When we go out for dinner with friends, the main object is not always to glean useful information but simply to bond with the people we are with and to enjoy their company.

A. Wallwork, *Meetings, Negotiations, and Socializing,*
Guides to Professional English, DOI 10.1007/978-1-4939-0632-1_13,
© Springer Science+Business Media New York 2014

13.2 Why do I find listening to spoken English so difficult but reading so easy?

When you read a text, the punctuation (commas, full stops, capital letters etc) helps you to move within a sentence and from one sentence to the next. Brackets, for example, show you that something is an example or of secondary importance. Punctuation also helps you to skim through the text without having to read or understand every single word. You don't really need to read every single word as you can recognize certain patterns and you can often predict what the next phrase is going to say.

A similar process takes place when you listen to someone speaking your native language. You don't need to concentrate on every word they say. Unfortunately, although we can usually quite easily transfer our reading skills from our own language into another, we cannot transfer our listening skills—particularly in the case of the English language. English often sounds like one long flow of sounds and it is difficult to hear the separations between one word and the next.

However spoken English does follow some regular patterns, and if you can recognize these patterns it may help you to understand more of what you hear and enable you to understand the general meaning rather than trying to focus on individual words and then getting lost!

In the spoken language, we often begin phrases and project our intonation in a particular way, but then we may abandon what we are saying – even in the middle of a word. Thus, unlike the written language, which generally has some logical sequence, the spoken language often seems to follow no logical track and is therefore more difficult to understand. However by recognizing the intonation we can get a clearer idea of the 'direction' in which the speech is going.

13.3 Why does spoken English sound so different from the English I learned at school?

When spoken at high speed, English words seem to merge together to create one long noise. For example, the simple question: do you want to go and get something to eat? when spoken fast becomes: wannagetsomingteat?

The problem for you and other non native speakers is that you have probably learned to say the first version, and you will therefore be unable to clearly recognize the native version.

So if spoken English is similar to the noise of an express train, how can you possibly understand a native speaker? The key is not to try and differentiate between the sounds and words, but to focus only on those parts of the phrase that are said the loudest and with the most emphasis. When native speakers say *do you want to go and get something to eat* they don't give each word the same stress and the same volume.

English is what is known as a stress-timed language, which means it has a kind of in-built rhythm that native speakers follow. This may be one of the explanations for the success of rock music and rap sung in English.

The words that are given the most stress are generally those that have the most importance in the phrase: *do you* want *to go and* get *something to* eat.

If someone said to you: *want get eat* with an intonation that suggested they were asking you a question, you would not have too much difficulty in understanding their meaning. So what you need to try and train yourself to do is just to focus on those words with the most stress. By 'stress' I mean a combination of three factors: clarity, volume and length. Thus the key words in a phrase tend to be articulated more clearly, at a higher volume, and at a slower speed.

This means that the words that add little value to the phrase are said much less clearly, at a lower volume and considerably faster. However, given that these words tend not to give key information, you can just ignore them and still have a good chance of understanding the overall meaning of the phrase. Basically the time it takes to say *eat* will be approximately the same as it take to say *to go and*.

In theory this might sound very logical and even obvious. Of course in practice it is much harder. However with some self-training, a more relaxed approach, and more realistic expectations about what you are likely to understand, this method is certainly less stressful than anxiously trying to understand each individual word in a phrase. If you focus on individual words and sounds, you will soon get lost and lose track of the conversation. If, instead, you focus on every fifth or sixth word, or where the words that create the rhythm then you will be more able to keep up with the conversation.

13.4 Do all native English speakers understand each other?

You may have experienced a feeling of inferiority and inadequacy when talking to native speakers. You feel nervous. You feel stupid. You feel that they cannot participate on an equal level.

But native speakers do not always understand each other. As Oscar Wilde noted: "We have really everything in common with America nowadays, except, of course, the language." And another Irish playwright, George Bernard Shaw once said: "England and America are two countries separated by a common language".

I am English and frequently in my life I have had experiences when I fail to understand another native English speaker. I once spent 20 min in a taxi in Glasgow (Scotland) in which the taxi driver chatted away to me in a strong Glaswegian accent, and I understood absolutely nothing. When I am in London, I am often unable to understand a single word in the first 30 s of what a native Londoner says to me because they speak so fast.

In the English language world we accept that often we do not understand each other.

The solution for you as a non native speaker is to change perspective. Think about situations within your own country. Presumably there are people in your country who speak with a very different accent from yours, and may even use a slightly different vocabulary or dialect. If you do not understand what they say, I imagine that you do not feel any sense of inferiority—at the most you might feel a little embarrassed. In any case, you probably collaborate with each other to understand what you are saying.

So when you have to speak to native speakers, try to imagine that English is simply a rather obscure dialect of your own language! Your objective and that of your interlocutor's is to understand each other. You should both be on an equal level.

13.5 Is it OK to tell my interlocutor that I am having difficulty in understanding him / her?

A key factor in your ability to understand native speakers is letting them know that you are not a native speaker and thus your command of the language is not the same as theirs. The problem with a lot of people whose first language is English is that they often don't learn languages themselves. They thus have no idea of the difficulties that you might experience in trying to understand them. Also they are not aware that you might for instance have a good command of spoken English and written English but that your listening skills are much lower.

If you don't encourage the native speaker to speak clearly, then you will significantly reduce your understanding of what they say. This is certainly not a benefit for you, and is probably not good for them either.

Instead you need to make it immediately clear to the native speaker that you need him or her to speak slowly and clearly, make frequent mini summaries and be prepared for many interruptions for clarification on your part. You could say something like this:

> It would be great if you could speak really slowly and clearly, as my English listening skills are not very good. Thank you. And also please do not be offended if I frequently ask you for clarifications.

But the problem does not end there. Even if the native speaker acknowledges your difficulties, they are likely to forget these difficulties within 2 or 3 min, as they then become absorbed by what they are saying. This means that you frequently have to remind them to speak more slowly.

> I am sorry, but please could you speak more slowly.

13.6 How can I concentrate more when I am listening?

We speak at between 120 to 150 words per minute, but as listeners our brains can process between 400 to 800 words per minute. This means that we get distracted easily and starting thinking about other things. If you really want to improve your listening and thus to understand better what other people are saying to you in English, then you need to focus exclusively on what this person is saying. If you start thinking about your next question (or other things) you will quickly get distracted. Try to think at the same speed as your interlocutor rather than being constantly ahead of them.

You can improve your chances of hearing what your interlocutor wants you to hear if you:

- focus not just on the first part of what someone says but also on the last part (our tendency is to listen attentively at the beginning and then half listen towards the end)

- participate in a conversation with an open mind, i.e. you need to put aside any prejudices you have about, for instance, politics, ethics and religion

- decide that the topic being discussed is potentially interesting, rather than immediately deciding that it is of no interest to you and think about other things

- pay attention, rather than pretending to pay attention

- try not to get distracted by how your interlocutor speaks or any other mannerisms they may have

- focus not just on the facts that you are given, but how these facts are given, and what interpretation your interlocutor is giving them

13.7 How can I prepare for a conversation so that I am likely to understand more of it?

You will massively increase your chances of understanding a conversation in a social context, if you prepare vocabulary lists connected with the kind of topics that might come up in conversation. It is true an infinite number of topics could be discussed at a social dinner, but what is also certain is that some topics come up very frequently. These topics include:

- the meetings that you have had already or have arranged

- any social events that have been arranged in conjunction with your visit

- the weather

- the food

- latest technologies (cell phones and applications, PCs, etc)

If you learn as much vocabulary as possible connected with the above points, you will feel:

- more confident about talking, i.e. offering your opinions and responding to others

- more relaxed when listening

The result will be that you will be able to participate in the conversation actively, and thus have a more positive and rewarding experience.

Other topics which are typically covered in non-work situations include: family, work, education, sport, film, music, and the political and economic situation of one or more countries. Again, if you learn the words (meaning and pronunciation) associated with these topics you will be able to participate much more effectively.

14 WHAT TO DO IF YOU DON'T UNDERSTAND WHAT SOMEONE SAYS TO YOU

14.1 Identify the specific word that you did not understand

Avoid saying 'repeat please' if you don't understand what someone has said.

Instead it is much more beneficial both for you and your interlocutor if you precisely identify what particular word or part of the phrase you didn't understand.

Let's imagine that the part in italics in the sentences the first column below is the part said by your interlocutor that you did not understand. Your questions for clarification are shown in the second column.

YOUR INTERLOCUTOR SAYS THIS:	TO CLARIFY, YOU SAY THIS:
I thought the food was *delicious*.	You thought the food was *what* sorry?
	Sorry, *what* did you think the food was?
I made *a terrible mess* with my presentation.	You made *a what* sorry?
	Sorry, *what* did you do with your presentation?
Last night I went *to the mosque*.	You went *where* sorry?
	Where did you go, sorry?
I have just had an interesting conversation with the *sales* manager.	With *which* manager sorry.
	Sorry, who did you speak to?

In the examples above, there are two types of clarification questions:

I thought the food was *delicious*.

Type 1) You thought the food was *what* sorry?

Type 2) Sorry, *what* did you think the food was?

Type 1 is easy to use because you simply repeat the words immediately preceding the word that you didn't understand (in the example the repeated words are shown in italics).

A. Wallwork, *Meetings, Negotiations, and Socializing,*
Guides to Professional English, DOI 10.1007/978-1-4939-0632-1_14,
© Springer Science+Business Media New York 2014

14.1 Identify the specific word that you did not understand (cont.)

Type 2 is more complex because it entails using an auxiliary verb (*did* in the example). It also means that you have to remember to put the words in a question form.

So it is simpler to use Type 1, which you can further reduce to: *the food was what?*

Both Types 1 and 2 have the advantage that your interlocutor will only repeat the word that you didn't understand (*delicious* in this case). This means that you now have a far greater chance of actually hearing the word, as it will now be detached from its surrounding words.

If you don't understand the meaning you can say: *Sorry what does that mean?* Or repeat the word with a rising intonation to indicate that you don't understand: *delicious, sorry?*

Obviously you can only use this strategy of repeating words if you hear the words in the first place.

Here is an extract from a conversation. It highlights the strategies that Person A used to identify what Person B has said.

1. A   Could you tell her that the meeting has been put off until next Thursday.
2. B   The meeting has been what sorry?
3. A   It's been put off until next Thursday.
4. B   Sorry, 'put off'?
5. A   Postponed. It has been postponed till next Thursday.
6. B   So it's been delayed?
7. A   Yes, that's right.

The key problem in the above dialog was that the woman did not know the term *put off*. The first time the man says *put off* (line 1), the woman probably hears the following sound: *zbinputofftil*, i.e. one long sound where several words have been merged together. Basically she has no idea what was the man said between *meeting* and *next Thursday*. So her tactic is to say all the words she heard until the point where she stopped understanding (line 2). The man then simply repeats the second part of his initial sentence (line 3), but this time probably with more emphasis on *put off*. Because the man puts more emphasis on *put off*, the woman is now able to identify the part of the sentence that she had failed to identify before.

14.1 Identify the specific word that you did not understand (cont.)

Note how in line 4 the woman begins her sentence with *sorry*. If she had simply said 'Put off?' the man might have just answered 'Yes, put off' as he would think she was just asking for confirmation. Instead the word *sorry* tells the man that the woman is not familiar with the term *put off*. *Sorry* is very frequently used by native speakers when asking for clarification.

The man now uses a synonym for *put off* (postponed). The woman's final strategy (line 6) is to check, using her own words, that she has understood correctly the meaning of *postponed*. And finally in line 7, the man confirms that she has understood.

The woman never says *Repeat please*. If you say *repeat please*, your interlocutor will just repeat the whole phrase again and you will probably not understand any more than you did the first time. Basically the woman asks a series of increasingly specific questions to identify the exact part of the man's phrase that she did not understand.

The above dialog and strategies used are typical of the strategies used by native speakers too. If you use them with a native speaker, he or she will not think you are doing anything strange.

14.2 Identify the part of the phrase that you did not understand

The examples given below are useful when you are focusing not just on an individual word, but on a whole sequence of words. If you don't identify the precise part of the sentence, your interlocutor will just repeat everything probably exactly as before, but perhaps a bit louder. By identifying the part you did not understand there is a chance that the interlocutor will use different words and say that particular part more slowly. The second column also shows alternative forms of clarification

WORD/S MISUNDERSTOOD	CLARIFICATION QUESTION
It chucked it down yesterday so I couldn't summon up the courage to venture out.	Sorry, I didn't understand the *first part* of what you said. Sorry *what* did it do yesterday?
It chucked it down yesterday so I couldn't *summon up the courage* to venture out.	Sorry, I missed the part in the *middle*. Sorry *what* couldn't you do?
It chucked it down yesterday so I couldn't summon up the courage to *venture out*. Note: The sentence means: It was raining very hard yesterday and I didn't feel like going outside.	Sorry, I didn't catch the *last part*. Sorry, I didn't understand *the bit after 'courage'*. Sorry, I didn't understand *what* you didn't have the courage to do. The courage to do *what*, sorry?

14.3 Avoid confusion between similar sounding words

Some words sound very similar to each other and are frequently confused even by native speakers. Below are some examples of how to clarify certain pairs of words.

WORDS	POSSIBLE MISUNDERSTANDING	CLARIFICATION
Tuesday vs Thursday	We have scheduled the meeting for Tuesday.	That's Tuesday the sixth right?
13 vs 30	We need thirty copies.	That's thirty, three zero, right?
can vs can't	I can come to the meeting.	So you are saying that you <u>are</u> able to come to the meeting?
	I can't come to the meeting.	So you mean that you <u>not</u> able to attend the meeting? So you mean that you <u>cannot</u> attend?

In the first example, the secret is to combine the day of the week with its related date. This means that your interlocutor has two opportunities to verify that you have understood correctly. If you have misunderstood, your interlocutor can then say: *No, Thursday the eighth.*

The confusion in the second example happens with numbers from 13 to 19 and 30, 40, 50 etc. Using the correct stress can help: thirt<u>een</u> vs <u>thir</u>ty. However, particularly on the telephone, this subtle difference in pronunciation may not be heard. So the secret is to say the number as a word (e.g. *one hundred and fourteen*) and then to divide it up into digits (*that's one one four*). If you have misunderstood, your interlocutor can then say: *No, thirteen, one three.*

In the third example, the problem is increased if *can* is followed by a verb that begins with the letter T. Thus understanding the difference between *I can tell you* and *I can't tell you* is very difficult. There are also significant differences between the way native speakers pronounce the word *can't* – for example, in my pronunciation *can't* rhymes with *aren't*, but for others the vowel sound of 'a' is the same as in *and*. The solution is to replace *can* and *can't* with the verb *to be able to*. You also need to stress the *are* in the affirmative version, and the *not* in the negative version, as illustrated in the table. If you have misunderstood, your interlocutor can then say: *No, I <u>am</u> able to come* or *no, I am <u>not</u> able to come* (alternatively *I <u>cannot</u> attend*).

14.4 Make frequent summaries of what your interlocutor has said

One way to understand more of what your interlocutor says is to make frequent short summaries of what your interlocutor tells you. Your interlocutor will appreciate the interest you are showing. You can use phrases such as the following to begin your summary:

Can I just clarify what I think you are saying? You mean that ...

I just want to check that I am following you correctly. So you are saying that ...

Your listener will not interpret such clarifications as a lack of English comprehension skills on your part, but that like a native speaker you simply want an accurate understanding of what has been said.

Using this tactic means that you could turn a potentially embarrassing situation into something positive.

14.5 Dealing with colloquialisms

Like all languages, English is full of colloquialisms. Unfortunately, native English speakers may not necessarily be aware which expressions are colloquial and which are not.

Many colloquialisms are in the form of phrasal verbs. Here are some examples, the phrasal verb is highlighted in italics.

Mark, let me *float* this *by* you. = get your opinion on

I'd like you to *fill* me *in* about ... = give me info about

Could you *run* the main points *by* me once more. = list, explain

This should help to *iron out* any problems. = resolve

OK, that just about *wraps* everything *up*. = concludes

Phrasal verbs tend to be made up of a monosyllable infinitive (e.g. *go, come, fill, run*) plus a preposition or adverb (e.g. *by, in, up, with*). Because of the shortness of the words, it is often difficult to understand them. An additional difficulty is that often a noun or pronoun splits the infinitive from the preposition / adverb.

There are thousands of phrasal verbs, so you cannot learn them all. The secret is simply to ask the speaker to rephrase or clarify what they have just said (as outlined in the subsections above).

14.6   Business jargon

While preparing for a meeting and negotiation, it is useful to learn some of the words and expressions that might come up during the discussion and which you might not have encountered before when learning and practising your English.

Learning such jargon is slightly easier than the phrasal verbs (see 14.5), because there is a limited quantity and the expressions are more intuitive (i.e. the combination of words generally helps you to understand the meaning).

Here are a few examples, the key word / expression is in italics.

We want to achieve a *win win situation*. = a situation in which both parties are happy

Can we just check that we *are on the same page*? = we agree with each other

Just to give you a *ballpark figure*. = an approximate number

We are not trying to *reinvent the wheel*. = to spend an unnecessarily long time on something

Let's *touch base* early next week. = contact each other to find out the current situation.

You can find more examples of jargon at: www.theofficelife.com/business-jargon-dictionary-A.html

Index

15.1 Making arrangements via email for meetings and teleconferences

Suggesting the time

Responding positively

Informing of unavailability at that time

Making an alternative suggestion

Changing the time

Confirming the time

Responding to confirmation of the time

Cancelling a meeting set up by the other person

Asking for directions to the meeting place

Arranging another meeting when first meeting has been concluded

15.2 Chairing a formal meeting

Checking everyone is present and has the agenda

Introducing people

Outlining the purpose of the meeting

Moving on to the next point

Deciding what to discuss next

Asking participants to focus

Countering an interruption

Summarizing

A. Wallwork, *Meetings, Negotiations, and Socializing,*
Guides to Professional English, DOI 10.1007/978-1-4939-0632-1_15,
© Springer Science+Business Media New York 2014

Taking a vote

Nearing the conclusion of the meeting

Ending the meeting

Asking someone to stay behind

15.3   Chairing an informal project progress meeting

Talking about tasks in the near future

Asking individuals about their work in progress

Telling individuals what they're doing and have to do in the future

15.4   Negotiating

Describing the company

Setting out aims and conditions

Discussing terms, conditions and dispute resolution

Offering extra items, services and discounts

Asking for more details

Showing that you are happy with the arrangements

Showing that you are unhappy with the arrangements

Summarizing what has been decided

Concluding the negotiation and talking about the next steps

15.5   Asking for and giving opinions, suggestions etc

Asking everyone's opinion

Asking for a specific person's opinion

Asking for reactions

Requesting suggestions and ideas

Giving opinions

Making suggestions

Making tentative suggestions

Making strong suggestions

Agreeing

Polite but strong rejection

Diplomatic disagreement

Picking up on what someone else has said

Interrupting your interlocutor

Questioning relevancy of what someone has just said

What to say when someone interrupts you

Returning to what you were just saying before an interruption

Returning to main point after an interruption (e.g. a phone call)

Beginning a parenthesis

Pausing for time

15.6   Checking understanding and clarifying

Asking the speaker to change their way of speaking

Reminding speaker to change their way of speaking

Asking for repetition of the whole phrase

Identifying the part of the phrase that you did not understand

Repeating the part of the phrase up to the point where you stopped understanding

When the speaker has repeated what they said but you still cannot understand

When you understand the words but not the general sense

When you didn't hear because you were distracted

Clarifying by summarizing what other person has said

Clarifying what you have said

Clarifying a misunderstanding in what you said

Clarifying a misunderstanding regarding what a third party has said

Checking that others are following you

Saying that you are following what someone is saying

Checking you have understood

Asking for clarification by repeating what they've said

Confirming that you understand

Underlining your main point

Losing track

15.7 Trade fairs

Initial introductions

Finding out about visitor to your stand

Explaining products and services

Suggesting times for further discussion

Customer: Explaining what you want

Apologizing for not being able to provide a particular product or service

Concluding the meeting

15.8 At the office

Introductions by guest

Arriving late

Welcoming guest at reception

Telling guest when he / she will be seen

Making small talk (receptionist and guest)

Giving and following directions within the office

Giving and following directions outside the office

Meeting people for the first time (previous contact via email, phone)

Meeting people for the first time (no previous contact)

Telling people how to address you

Introducing people

Meeting people who you think you may have met before

Seeing people you have already met before

Catching up

Finding out about a person's job

Describing your job

Receiving guests by person who has arranged meeting with the guest

Saying goodbye

Leaving reception: guest

15.9   Socializing

Finding out where someone comes from

Holidays

Guest questions to host

Family

Talking about language skills

Discussing differences between countries

Talking about food

Discussing politics

Someone who has moved to your country

Showing interest

Enquiring

Responding to an enquiry

Requesting help

Accepting request for help

Declining request for help

Offering help

Accepting offer of help

Declining offer of help

Giving advice

Showing enthusiasm

Giving condolences

Making excuses for leaving

Using the time as an excuse for leaving

Wishing well and saying goodbye

15.10   Traveling

Buying air tickets

Buying train tickets

Dealing with taxis

Giving directions

15.11   Hotels

Reserving a hotel room

Asking about hotel location and facilities

Arriving at hotel

Asking about services

Problems with the room

Checking out

15.12   Restaurants

Formal invitations for dinner

Accepting

Responding to an acceptance

Declining

Responding to a non-acceptance

Arriving at a restaurant

Menu

Explaining things on the menu and asking for clarification

Toasting

Making suggestions

Saying what you are planning to order

Requesting

Declining

Being a host and encouraging guests to start

Being a guest and commenting on food before beginning to eat

Asking about and making comments on the food

Ending the meal

Paying

Thanking

Replying to thanks

15.13   Bars

Suggesting going to the bar / cafe

Offering drink / food

Accepting offer

Mistakes with orders

Questions and answers at the bar / cafe

15.1 MAKING ARRANGEMENTS VIA EMAIL FOR MEETINGS AND TELECONFERENCES

Suggesting the time

Let's arrange a call so that we can discuss it further.

Can we arrange a conference call for 15.00 on Monday 21 October?

Would it be possible for us to meet on …?

How about Wednesday straight after lunch?

Could we meet some time next week?

When would be a good time?

What about December 13?

Would Friday at 4 o'clock suit you?

Would tomorrow morning at 9.00 suit you?

Shall we say 2.30, then?

Responding positively

OK, that sounds like a good idea.

Yes, that's fine.

Yes, that'll be fine.

That's no problem.

Informing of unavailability at that time

No, sorry. I can't make it then.

I'm afraid I can't come on that day.

Sorry but I can't make it that day.

Sorry but I'll be on holiday then.

I'm afraid I have another engagement on 22 April.

Would love to meet – but not this week! I can manage Nov 16 or 17, if either of those would suit you.

I am afraid next week is out because …

Making an alternative suggestion

Tomorrow would be better for me.

If it's OK with you, I think I'd prefer to make it 3.30.

Could we make it a little later?

Could you make it in the afternoon?

Could you manage the day after tomorrow?

Changing the time

Sorry, I am afraid I can't make the meeting at 13.00. Can we change it to 14.00? Let me know.

Re our meeting next week. I am afraid something has come up and I need to change the time. Would it be possible on Tuesday 13 at 15.00?

Unfortunately, I'll have to cancel our meeting on ...

I'll be unable to make the meeting.

We were due to meet next Tuesday afternoon. Is there any chance I could move it until later in the week? Weds or Thurs perhaps?

Could we fix an alternative?

Can we fix a new time? How about ...?

Confirming the time

The meeting is confirmed for Friday at 10:30 am Pacific time, 12:30 pm central time. Please send any items you want to discuss, and I will send an agenda earlier in the morning.

Responding to confirmation of the time

I look forward to seeing you on 30 November.

OK, Wednesday, March 10 at 11.00. I look forward to seeing you then.

OK, I will let the others know.

Cancelling a meeting set up by the other person

Something has come up, so I'm afraid I can't come.

Sorry but the other members of my group have arranged for me to ...

Sorry but it looks as though I am going to be busy all tomorrow. The thing is I have to ...

Asking for/giving directions to the meeting place

Can you give me some directions?

I would take a taxi from the airport.

Will you be coming by car?

I'll email you a Google map.

Just ask for me at reception.

My office is on the third floor.

Arranging another meeting when first meeting has been concluded

It might be a good idea if we could arrange another meeting to discuss a few things in more detail.

I don't know if we'll be able to finish everything today.

It might not be a bad idea if we could arrange another meeting.

Could you meet up again say at the end of the month?

Perhaps we could see each other again next Monday, as I'll be back in *country / town* then anyway.

Would there be any chance of you coming to us this time?

I've got a really busy schedule that week and I really won't have time to fly over.

I'm sorry I haven't got my diary with me.

The best thing would be if I could ring you tomorrow morning, if that's OK with you.

What about Tuesday at 08.30, or is that too early for you?

No, that would be fine. Tuesday at 08.30 then.

15.2 CHAIRING A FORMAL MEETING

Checking everyone is present and has the agenda

Is everybody here?

I think we are all here apart from …

Did everyone get my email with the agenda?

OK, we'll just wait another couple of minutes and then we need to start.

OK, so I think we are all here now, so let's start.

Introducing people

For those who don't know me, I am …

On my right is *name*, who will be taking the minutes of the meeting.

On my left is *name*, who is responsible for …

Can I just ask your names and positions?

Outlining the purpose of the meeting

I've called this meeting to discuss …

This meeting has been arranged in order to ...

The main purpose of this meeting is ...

Other objectives of this meeting are to ...

We have just under one hour for the meeting.

I plan to end this meeting at not later than 11.00.

Moving on to the next point

Let's move on to the second point now.

Shall we continue then?

Why don't we move on to the next point?

Deciding what to discuss next

Shall we hear the figures now?

Let's discuss the results first.

I suggest we postpone the decision till the next meeting.

Up to a point, I agree with you, but ...

Asking participants to focus

Can we focus on the matter in hand?

Can we just do x and then go back to y later?

I think we're losing sight of what we are trying to do so can we move on to ...?

Countering an interruption

Could we just let Stefan finish?

Pietro – you were saying?

Sorry, were you about to say something?

Summarizing

So, basically what we're saying / proposing is.

In conclusion ...

To sum up ...

So, if you'd like me to summarise what we've ...

So just to summarize what we've been saying ...

So, drawing together what we've said ...

First we've got to …

Then, I think we really need to …

Finally, I'd like you to …

Are we all clear about what we have agreed to do?

OK, thanks for coming.

Taking a vote

Can we just have a show of hands for those in favor?

And for those against?

OK, well it looks like the motion has been passed.

Nearing the conclusion of the meeting

We've only got a few minutes left.

Anything else anyone wants to add?

We're running (we've run) out of time.

Shall we call it a day?

Shall we wind things up?

Ending the meeting

I think we've covered everything so let's finish here.

OK, I've said all I want to say, so unless any of you have anything to add, we can stop here.

This is a good point to end the meeting.

Right, that just about wraps things up.

Asking someone to stay behind

Pietro, before you go, do you think I could have a quick word. Thanks.

15.3 CHAIRING AN INFORMAL PROJECT PROGRESS MEETING

For phrases on how to open and conclude a meeting, see previous section.

Talking about tasks in the near future

I think in the next few days we need to have finished doing x.

We really need to do this before the end of the week.

Marcus is away this week, so I need someone to cover his work.

We've got to give Stuart feedback by Friday. But even if you have no comments to make, please send a mail in any case.

I think that by Wednesday we should be able to make the release.

I would like to have x ready by Thursday.

Please could we have x ready by Thursday.

Some of you will be having a meeting with x on Friday.

Please make sure that …

Next week we're going to start working on x again.

Let's see if we can …

Asking individuals about their work in progress

Melanie, you're supposed to be doing Y. How are you getting on?

Daniel, have you had time to do Z?

Curtis could you get feedback from Q? Have you had any feedback from X?

Monique, where are we up to with P?

Sushil, you're responsible for doing X, how's that going?

Telling individuals what they're doing and have to do in the future

Kasper, you're looking after X. I think you've got a release to make by the end of tomorrow.

Alexander, I need you to do X. Could you get me Y within the next couple of days please.

I'd like to be able to have completed the second document by Thursday.

Please could we be sure to have done X by Friday.

It would be good if you could be proactive on this.

Let's try and do our best on this.

I'd like to see X done by Thursday.

15.4   NEGOTIATING

Describing the company

The company was founded in 2011.

We set up the company in 2011.

The company was first registered in 2001.

The company was taken over by ABC in 2013.

In 2014 we merged with …

We went public 2 years ago.

There are four partners in the firm.

We employ 10,000 people.

We have 10,000 employees.

Our main markets are …

Traditionally, our main market sector has been …

We are now seeing a large growth in …

We have an annual sales / turnover of $250 million.

Annual profits are in the region of £ 25 million.

The net profits this year amount to … compared with last year.

During the year the demand for … went up / down considerably / slightly.

A new subsidiary has been formed to handle the company's IT plans.

We are planning to float the company next year.

Setting out aims and conditions

Essentially, what we are looking for is …

Ideally, what we would like to do is …

Basically, we are interested in …

We have four basic requirements which are not really open to negotiation.

We are however prepared / happy to negotiate the terms of the …

If you are prepared to do x, then we would be prepared to do y.

If you think that x would make a difference, then we could try to …

Our position is that we …

Discussing terms, conditions and dispute resolution

While we understand that you … I am afraid we cannot …

Unless you are prepared to do x, then I am afraid we cannot do y.

Would you be willing to …?

Will you be able to guarantee …?

If we were to do x, would you then be willing to do y?

If you need to discuss this amongst yourselves or make a phone call, then perhaps we can have a break for 20 minutes.

If we accept your prices, then we will have to raise our prices.

If we agreed to that it would not be good for our business.

If you can reduce your price by ..., then we will ...

We must insist on delivery within the time stated and reserve the right to reject the goods should they be delivered later.

What happens if there is a dispute?

What compensation will you pay if ...?

The contract would have to be governed by the jurisdiction of our country.

Offering extra items, services and discounts

In addition, we/you can deliver the goods on 25 September.

We can supply the products by 25 September.

Our lines are mainly ...

We can offer a large variety of ...

We are able to quote you very advantageous terms.

Asking for more details

Can you arrange delivery to our site by truck?

What discount / warranties / guarantees can you offer?

What are the terms of your license agreement?

On what are those figures based?

Showing that you are happy with the arrangements

We accept / agree.

That's great.

OK, I think we have a deal.

Showing that you are unhappy with the arrangements

I am afraid we cannot change our offer.

This is our final offer.

Unfortunately we cannot accept your payment/delivery/discount terms.

Summarizing what has been decided

We have covered a lot of ground in this meeting.

Let me go over all the details again.

Have I / we covered everything?

Are there any questions?

Concluding the negotiation and talking about the next steps

Do you accept these terms?

Can you prepare a draft contract?

I will draft an outline agreement.

I will email you the agreement for your comments.

Could you kindly email me the draft contract for our comments.

15.5 ASKING FOR AND GIVING OPINIONS, SUGGESTIONS ETC

Asking everyone's opinion

Do you all agree on that?

Does anyone have any comments?

What do you think about the budget?

What are your feelings about the budget?

What are your views on this?

What's your opinion?

How do you see this?

Asking for a specific person's opinion

Pietro – what do you think about ....?

Clara, would you like to comment here?

Teresa, what about you?

Pete, can I just bring you in here?

Asking for reactions

Any reaction to that?

What's your reaction to that?

Has anybody any strong feelings about that?

Has anybody any comments to make?

Requesting suggestions and ideas

Any suggestions?

Do you think we should …?

I'd like to hear your ideas on this

Do any of you have any suggestions?

How do you think we should do this?

What would you recommend?

Do you think we should …?

I suggest we should … What do you think?

Giving opinions

I think / reckon we should …

What I think is …

I honestly think that …

The way I see it …

It seems to me that …

As I see it ….

My inclination would be to …

From a financial point of view …

I tend to favor the view that …

Making suggestions

What about …?

Why don't we …?

What I think we should do is …

I (would) recommend / suggest that we should …

My recommendation is that we should …

We should / ought to …

If I were you I would …

Making tentative suggestions

We could always …

It might be a good idea to …

Have you thought of …?

One solution would be to …

What about …?

Is there any reason why we shouldn't …?

I wonder if we could …

What would it look like if we…?

Making strong suggestions

The only solution is to …

I see no other alternative but to …

There is no alternative but to …

We must …

Agreeing

I'm in complete agreement.

I couldn't agree more.

I (quite) agree.

Right.

You're right there.

I think you're right.

Yes, definitely.

Exactly!

Precisely!

Polite but strong rejection

I'm afraid I can't accept that.

I'm sorry, but that's not really practical.

I'm afraid I'm not very happy about that.

I'm sorry, but I have reservations about that.

I really think we should concentrate on X. I don't want to put it in the background.

I don't want to force the issue more than necessary, but ...

I suppose so, but I still think ...

Actually, I'm not sure that that is necessarily the best approach.

Diplomatic disagreement

I see what you mean, but ...

You've got a point, but ...

I take / see your point but ...

I appreciate what he's saying but ...

I appreciate your point of view but ...

You may be right, but personally I ...

I'm not sure whether that's feasible ...

I don't want to sound discouraging but ...

I accept the need for x but ...

I can see why you want to do this but ...

OK, but what if ...?

Yes, but have you thought about ...?

Picking up on what someone else has said

On that subject, I think ...

As far as the budget is concerned I think ...

Regarding the budget, I think ...

While we're on that subject, I think ...

With regard to the budget, I reckon ...

Interrupting your interlocutor

Excuse me for interrupting.

May I come in here?

I'd like to comment on that.

If I could just interrupt you ...

OK, but listen ...

Sorry, could I just interrupt?

Sorry do you mind if I just say something?

Sorry do you mind if I just ask Luigi a question?

We're talking at cross purposes.

Questioning relevancy of what someone has just said

That's not really the point.

I'm not sure that's really relevant.

I'm not sure what that's got to do with it.

What to say when someone interrupts you

Sorry, just a sec ...

OK, I've nearly finished ...

Sorry, if I could just finish what I'm saying ...

Can I just finish what I was saying? It will only take me a minute.

Sorry, just one more thing, ...

Sorry, can I just say / add something.

I would just like to add that ...

Returning to what you were just saying before an interruption

As I was saying ...

Going back to what I was saying / I said before ...

Let's just go back a bit to what we were saying before.

Can I just go back ...

Let's get back to the point.

I think we're losing sight of the main point.

Returning to main point after an interruption (e.g. a phone call)

OK, where was I? / What was I going to say?

OK, what we were saying? Oh, yes, I was saying that ...

Beginning a parenthesis

By the way, did you know that Silvia is ...

By the way, I forgot to tell you that ...

On a completely different subject ...

If I could just change the subject a second ...

Pausing for time

I mean. Well. Right. Um. Er. You know.

Could we come back to that later?

Now where was I?

Sorry, I'll just have to think about that a sec.

Sorry, I've forgotten what I was going to say.

15.6   CHECKING UNDERSTANDING AND CLARIFYING

Asking the speaker to change their way of speaking

Sorry, could you speak up please?

Sorry, could you speak more slowly please?

You'll have to speak more slowly, sorry.

I don't want to sound rude but could you speak more clearly please?

Reminding speaker to change their way of speaking

Sorry, I really need you to speak up please.

Sorry, my listening skills are not very good, would you mind speaking more slowly please?

Sorry, my English is not very good, could you speak very slowly please.

Asking for repetition of the whole phrase

I'm sorry what did you say?

Could you explain that again using different words?

Sorry, could you say that again?

Sorry, I didn't catch that.

Sorry what was your question?

Identifying the part of the phrase that you did not understand

Sorry, what did you say at the beginning?

I didn't get the middle / last bit.

Sorry what was the last bit?

Could you say that last bit again?

Sorry I missed the bit about …

And you did what sorry?

And you went where sorry?

You spoke to who sorry?

Repeating the part of the phrase up to the point where you stopped understanding

Sorry, you thought the presentation was … ?

And then you went to … ?

And the food was …?

When the speaker has repeated what they said but you still cannot understand

Sorry, I still don't understand.

Sorry, do you think you could say that in another way?

Sorry, could you that again but much more slowly?

Sorry, could you write that word down, I can't really understand it.

When you understand the words but not the general sense

Sorry, I'm not really clear what you're saying.

Sorry I think I have missed the point.

Sorry but I am not really clear about …

When you didn't hear because you were distracted

Sorry, I missed that last part.

Sorry, I got distracted. What were you saying?

Sorry, I've lost track of what you were saying.

Sorry, I've forgotten the first point you made.

Sorry, I'm a bit lost.

Sorry I wasn't concentrating, what were you saying?

Clarifying by summarizing what other person has said

So what you're saying is …

So you're saying that it *is* true.

So if I understood you correctly, you mean …

Let me see if I have the big picture. You're saying that …

Clarifying what you have said

What I said / meant was …

What I'm trying to say is …

The point I'm making is …

Let me say that in another way.

In other words, what I mean is …

Clarifying a misunderstanding in what you said

No, that's not really what I meant.

No, actually what I meant was …

Well, not exactly.

What I was trying to say was …

That's not actually what I was trying to say.

Clarifying a misunderstanding regarding what a third party has said

I think you may have misunderstood what he said. What he meant was.

No, I think what he was trying to say was … Have I got that right?

If I'm not mistaken, what she was saying was:

Checking that others are following you

Does that make sense to you?

Do you understand what I mean?

Am I making myself clear?

Are you with me?

Do you see what I mean?

Are you following me?

Does that seem to make sense (to you)?

Do you understand what I'm saying?

Saying that you are following what someone is saying

Yes, I see what you're getting at.

Yes, perfectly.

Yes, I know what you are saying …

Yeah, yeah, yeah – I've got you.

I'm with you.

OK, I think it's clear what you are saying.

Checking you have understood

Let me check that I've understood.

I'm not sure I understand. Are you saying that ...?

Before you go on, do you mean that?

It is still not clear to me. What do I do when ...?

Let me check that I've understood.

I'm not sure I understand. Are you saying that ...?

Sorry but I am not really clear about ...

Sorry I wasn't concentrating / got distracted / was daydreaming, what were you saying?

Asking for clarification by repeating what they've said

Before we go on let me paraphrase what I think you are saying.

Let me restate your last point to see if I understand.

So what you're saying is ...

So if I understood you correctly, you mean ...

Confirming that you understand

Yes, I see what you're getting at.

Yes and no, it seems a bit of a contradiction to me.

I go along with you when you say X, but not with Y.

Underlining your main point

What I'm trying to say is ...

The point I'm trying to make is ...

Basically what I'm saying is ...

The thing is ...

To cut a long story short ...

All I'm saying is ...

Losing track

Sorry, I've lost track of what I was saying.

Sorry, I can't remember what I wanted to say / I was going to say.

Sorry, where was I?

Sorry, I just can't think of the word.

Sorry, can we come back to this later?

Sorry, I've lost track of what you're saying.

Sorry I think I am missing / have missed the point.

Sorry, I've forgotten what I was going to say.

Sorry, I can see I'm not making much sense.

15.7 TRADE FAIRS

Initial introductions

Good morning / afternoon. My name's …

Here's my card.

Do you know our company?

I'm sure you know our organization.

We make / provide …

We are the largest company in our field.

Finding out about visitor to your stand

What do you do?

Who do you work for?

What line of business are you in?

What does your company do?

What exactly would you like to know?

Explaining products and services

Would you like some information about our product / service?

Any specific product / service?

We have a number of products which can …

For your needs, I would recommend …

This product would be perfect for your needs.

This service exactly covers your needs.

This software could be the answer.

This equipment may be able to solve your problem.

Suggesting times for further discussion

We would be very pleased to …

visit your company.

prepare an estimate.

discuss this over dinner.

Would you like …

someone to visit your company?

us to prepare a quotation?

to discuss this over lunch?

to return to the stand / booth later?

a demonstration of the equipment?

Please take a copy of our brochure. It contains all the product information.

Customer: Explaining what you want

I am interested in …

I would like to know more about …

Could you explain exactly …

how this machine works?

what this product does?

where this service is provided?

if this service is available here?

We would like …

someone to visit our company.

to have a quotation.

to discuss this further.

to see a demonstration.

Apologizing for not being able to provide a particular product or service

I'm afraid we don't make anything like that.

I'm sorry but we don't provide that service.

Concluding the meeting

OK, so we'll see you again here at the stand at 9.00 tomorrow morning.

OK, so I will expect a call from you next week.

We'll be in touch with you next week / month.

Goodbye.

15.8   AT THE OFFICE

Introductions by guest

My name's ...

I've got an appointment with ...

Where can I leave my luggage?

Can you tell me where the toilet is?

Could you tell me where the bathroom is?

Arriving late

I'm sorry I'm late, but the taxi driver got lost getting here.

... but my plane was delayed.

... but I couldn't find your office.

... but the hotel forgot to give me my early morning call.

... but I got stuck in the traffic.

Could you ring him and apologise for me that I'm late.

Welcoming guest at reception

Good morning. You must be Mr Y, we've been expecting you.

Could you give me your name please?

Could you just sign the register please?

And you'll need one of these badges.

I'll tell Mrs Z that you're here.

Please take a seat.

I'll try to find someone who can help you.

Excuse me, but I didn't catch your name.

If you could wait a moment, please.

If you would like to sit down there.

Telling guest when he / she will be seen

Would you like to sit down, Mr X will be with you in a second.

She'll be with you shortly.

Mr X will see you now.

I'm sorry but Mr X won't be here – something unexpected has happened.

I'm really sorry but Mr X's been held up.

… has been delayed.

He'll be here in half an hour.

Something has cropped up.

I'm sorry but Mr. X is still held up in a meeting.

… with a client.

He'll be able to see you in half an hour.

Mr X is ready to see you now.

Mr X has asked me to ask if you'd mind waiting for five minutes.

Is there anyone else you'd like to see while you're waiting for Ms Y?

Can I get you / would you like a coffee?

Would you like something to read while you're waiting?

Have you seen our new brochure?

If you need me I'll be in the next room.

The bathroom's the first on your left.

If you'd like to make a phone call, you can use the one in the next room. Remember to dial zero to dial out.

Making small talk (receptionist and guest)

What part of *country* do you come from?

Whereabouts exactly?

Is this your first trip to *country*?

Will you have time for any sightseeing?

Isn't it a lovely day? I bet it's not like this in *country*!

Isn't the weather terrible?I suppose it was the same in *country*?

Giving and following directions within the office

Follow me, please.

Would you like to follow me, please?

This way please.

Here's the lift.

After you.

You first.

I prefer to take the stairs.

I'm sorry I can't take you up to his office, I've got to stay here on reception.

If you could just take the lift up to the third floor and Mr X's secretary will be waiting for you there.

Go down this corridor then down the stairs, at the bottom turn left and then left again and in front of you you'll see the bathroom.

Go to the top of those stairs over there. Turn right and the second door on your left is Mr X's.

The stairs are just round the corner.

Giving and following directions outside the office

Can you recommend anywhere where I could go for an hour or so?

Is it in walking distance?

How long will it take me to get there?

I'll just take you outside because it'll be easier to explain how to get there.

Go straight on until you get to the river.

Cross the river.

Then carry on straight for about five minutes.

When you get to the end of the road turn left.

And you'll see the tower at the end of that road.

It should take you about 15 minutes.

Meeting people for the first time (previous contact via email, phone)

Hello, pleased to meet you finally.

So, finally, we meet.

I'm very glad to have the opportunity to speak to you in person.

I think we have exchanged a few emails, and maybe spoken on the phone.

Meeting people for the first time (no previous contact)

Good morning, I'm …

Hello, I don't think we've met. I'm …

Pleased to meet you.

Nice to meet you, too

May I introduce myself? My name is …

I'm responsible for / I'm in charge of…

I'm head of…

How do you do?

Here is my card.

Do you have a card?

Telling people how to address you

Please call me Holger.

OK, and I'm Damo.

Fine, please call me Damo.

Everybody calls me …

Introducing people

May I introduce you to …

Have you two met?

Can I introduce a colleague of mine? This is Irmin Schmidt.

Hello, Pete, this is Ursula.

David, this is Olga. Olga, this is David.

I'm afraid Wolfgang cannot be with us today.

Meeting people who you think you may have met before

Excuse me, I think we may have met before, I'm …

Hi, have we met before?

Hi, you must be …

Seeing people you have already met before

Hi, Tom, good to see you again, how are you doing?

Hi, how's it going? I haven't seen you for ages.

How's things?

Great to see you.

I'm (very) pleased to see you again.

Catching up

How did the trip to Africa go?

How's the new job going?

How's your husband? And the children?

How is the new project going?

Finding about a person's job

What exactly do you do at ABC?

How long have you been with ABC?

When did you join the company?

What position did you start off at with ABC?

Who did you work for before?

Do you plan to stay with them?

What do you like most about your job?

Describing your job

I'm the Chief Technical Officer.

I'm in charge of sales in China.

I'm a senior / junior developer.

I am an engineer.

I deal with …

I am responsible for …

I am in charge of …

So, you've worked there nearly all your life?

How long have you been working for your company?

Have you been away much on business recently?

Receiving guests when you have arranged the meeting

Pleased to meet you.

Please sit down.

Pleased to meet you, My name's ...

How do you do? (I'm ...)

I've just got to make a quick phone call, and then I'll be with you.

If you wouldn't mind waiting a minute, I've just got to get this letter finished.

Did the secretary get you a coffee?

Did you have a good journey?

Did you have any difficulty getting here?

Which hotel are you staying at?

Did you arrive this morning or last night?

Would you like a coffee first?

Or shall we get straight down to business?

We've got 20 minutes before the meeting starts, shall we just go down to the bar and get a coffee?

Saying goodbye

I'm afraid but I've got to go.

I'm sorry but it's time for me to go.

Well, it's been a very useful meeting, thank you so much for coming.

Thank you for finding the time for this meeting.

I hope to be able to meet you again in the near future.

It was a pleasure to meet you.

Please send my regards to Mr X.

Please say hello to Ms Y.

See you next week in *place* then.

Hope to see you before too long.

Leaving reception: guest

I left my case this morning.

No, not that one.

Yes, that's it.

Could you ring a taxi for me?

Well, that's everything. Thank you for your help.

My taxi's here.

Thank you.

15.9 SOCIALIZING

Finding out where someone comes from

So where are you from exactly?

So are you from NY?

No, I'm from Poland

Ah from Poland?

Oh really, so what brought you to NY?

So how do you find NY?

Whereabouts is *place*? Where exactly is *place*?

How big is it?

I've heard there is a famous building in *place*, but I can't remember the name?

Is it true that it's famous for windsurfing? How come?

What is the weather like there?

Holidays

Have you taken any holiday yet this year?

Where did you go last year?

Had you been there before?

Guest questions to host

Do you live anywhere near here?

How long does it take you to get to work?

Do you come by car?

Family

Have you got any children?

Yes, two boys and a girl.

One daughter / three sons.

How old are they?

They're ten and twelve.

The oldest / youngest is …

Oh, they've left home.

Talking about language skills

How and where did you learn English?

Does your husband / wife speak English?

How many languages do you speak fluently?

Do you speak any other languages?

How difficult is your language to learn for a foreigner?

For me it's quite difficult to express emotions or complex ideas in another language, do you find this too?

Discussing differences between countries

The examples below refer to questions that an Italian asks a Spaniard with regard to differences between Italy and Spain. However, the questions could be adapted so as to refer to any country.

What do you think are the main differences between Italy and Spain?

Do you find the Italians are similar to the Spanish?

How similar do you think Italy and Spain are?

What are the differences between our life style and yours?

What are the main differences between the Spanish and Italian way of life?

Are there many differences between the way Spanish men and Italian men behave?

What do the Spanish think about the Italians?

I am going to be moving to Spain quite soon – do you think it's a nice place to live?

Talking about food

What do you think of the food here? How does it compare with food in your country?

Have you tried any of the local dishes?

Do you have any particular national / local dishes in your country?

What do you normally have for lunch in your country?

What do you miss about food from your own country?

Have you been to any restaurants here? How do the prices compare with restaurants in your country?

Have you tried the wine here?

Do people drink alcohol in your country?

Discussing politics

Politics can be a very dangerous topic, so be very careful when asking the questions below.

I'm curious to know if people in your country approve of your [left / right-wing] government?

What is the situation with x at the moment?

What does the general public / do the people think about x?

I have heard that your government has banned x – what reaction did this have? How has this impacted on life in your country?

What is the attitude to ...?

Do you know anything about politics in my country?

What do you think about our politicians? And our current government?

What does the press in your country say about our ...?

What do you think about our public services (schools, national health system, transport)?

Someone who has moved to your country

Note: The following phrases are related to questions to someone who is living in Vietnam but is not native to Vietnam. Obviously you could substitute Vietnam / Vietnamese for whatever country you wish to talk about.

Why did you decide to move to Vietnam?

How long have you been living here?

Had you ever been to Vietnam before?

Where did you live before coming to Vietnam?

How are you managing here in Vietnam? Are you learning much Vietnamese?

What have you seen in Vietnam?

Where have you been in Vietnam?

What do you think about Vietnam? And the life here?

What in particular do you like about Vietnam?

What do you think of the Vietnamese?

Have you made friends easily?

Are you going to stay here for a long time?

Do you have any particular plans for the future?

Are you planning on staying here or where else would you like to live?

What do you miss most about your home country?

Showing interest

Oh, are you?

Oh, is it?

Oh, really?

Right.

That's interesting.

Oh, I hadn't realized.

Enquiring

I wonder if you could help me?

Do you know where / how I could … ?

Do you happen to know if … ?

Excuse me, do you think you could … ?

Responding to an enquiry

Yes, of course.

Certainly. Sure. Yes, what's the problem?

No, I'm sorry I don't actually.

I don't actually, but if you ask that man …

Actually, I can't I'm afraid.

Requesting help

Do you think could you give me a hand with …?

Would you mind helping me with ...?

I wonder if you could help me with ...?

Could you give me some help?

Could you do me a favor?

Would you like me to give you a hand with ...?

Accepting request for help

Sure. No problem.

Two seconds and I'll be with you.

OK. Right. Where shall I start?

Declining request for help

I'm sorry but I can't just at the moment.

Sorry, but you've caught me at a bad time.

Offering help

Shall I help you with ...?

Do you want me to help you with ...?

If you want, I could give you hand with that.

Are you sure you don't need any help with that?

Accepting offer of help

That's really kind of you.

Great thanks.

If you're sure you can spare the time, that'd be great.

If you really don't mind, that'd be most helpful.

Declining offer of help

That's very kind of you but I think I can manage.

No, it's alright thanks.

Thanks but I really don't want to put you out.

Giving advice

Have you thought about ... ?

Don't you think perhaps you should ... ?

Perhaps it might not be a bad idea to ...

If I were you I'd ...

Maybe the best thing would be to …

Perhaps you ought to/should …

Showing enthusiasm

That's wonderful/great/fantastic/perfect.

Well done! Congratulations! Good on you.

That's marvelous news. I'm so pleased for you.

Really? I can hardly believe it.

You must be so proud of yourself.

Giving condolences

Oh well, it's better than nothing.

Bad luck! Better luck next time.

Oh dear! I'm sorry to hear that.

Well, I'm sure you did everything you could.

Making excuses for leaving

I am sorry – do you know where the bathroom is?

It was nice meeting you but sorry I just need to go to the bathroom (GB) / restroom (US).

Sorry but I just need to answer this call / I have just remembered I need to make an urgent call.

It has been great talking to you, but I just need to make a phone call.

Sorry, I've just seen someone I know.

Sorry, but someone is waiting for me.

Listen, it has been very interesting talking to you but unfortunately I have to go… may be we could catch up with each other tomorrow.

Using the time as an excuse for leaving

Does anyone have the correct time because I think I need to be going?

Oh, is that the time? I'm sorry but I have to go now.

Sorry, I've got to go now.

I think it's time I made a move.

Wishing well and saying goodbye

Formal

It's been very nice talking to you.

I hope to see you again soon.

I really must be getting back.

I do hope you have a good trip.

It was a pleasure to meet you.

Please send my regards to Dr Hallamabas.

Informal

Be seeing you.

Bye for now.

Keep in touch.

Look after yourself.

Say 'hello' to Kate for me.

See you soon.

See you later.

Take care.

See you in March at the conference then.

Hope to see you before too long.

Have a safe trip home.

OK, my taxi's here.

Nice to have met you.

Give my regards to Julia.

Say hello to Stefan.

Have a nice time!

Thanks very much, the same to you.

15.10 TRAVELING

Buying air tickets

Is there a flight to ...?

When does it leave / take off?

When does it arrive / land?

What time do I have to check in?

I'd like to book a return / round-trip flight to …

I'd like to cancel / change my reservation on flight number …

Buying train tickets

What's the fare to …?

Do I have to change?

When does it arrive at …?

Which platform does the train leave from / arrive at?

I'd like a single / one-way ticket to …

How much is a return / round-trip to *place* in first class?

I'd like to reserve a seat.

Dealing with taxis

Could you get me a taxi?

Where is the taxi rank / stand?

To the station please.

Please stop here.

Could you wait for me?

I'll be back in 10 minutes.

How much do I owe you?

Keep the change.

Giving directions

Go straight ahead.

It's on the left / right.

Opposite / behind …

Next to / After …

Turn left at the … next corner / traffic lights.

Take the A3.

You have to go back to …

15.11   HOTELS

Reserving a hotel room

Your hotel has been recommended to me by ...

Please could you reserve me a single / double) room from ...

How much is it per night, half board / full board, please?

Do you take credit cards?

Is breakfast included?

I'd like a single room for two nights.

I'd like a room with a shower.

I'll be arriving late.

How much does it cost?

Do you accept credit cards?

We would like to change our reservation at your hotel from *date* to *date*.

I would be grateful if you would confirm this booking.

Asking about hotel location and facilities

Could you please send us information about the hotel, its locations and the facilities you offer, plus details of your rates.

Could you please let us know where your hotel is located with respect to the centre of the town.

Could you send me a list of the agencies organizing guided tours in your region?

I'd like to know if you organize trips to *place*.

Do you organize any trips where it is possible to practice *sport*?

Arriving at hotel

My name's ...

I've got a reservation for two nights.

I have a reservation in the name of ...

The booking was confirmed both by email and fax.

Which floor is my room on?

When will it be ready?

Has anyone else from my company arrived here already?

I will be leaving at 08.30 tomorrow morning.

Asking about services

Is there an Internet connection?

Is there a shuttle bus to *place*?

Can you book me a taxi?

Is there a train that goes to the *place*?

What time do I have to be back at the hotel?

When is breakfast served?

I'm expecting Mr X at 7.00. Could you call me when he arrives?

Problems with the room

This key doesn't seem to work.

I have locked myself out.

My room has not been cleaned.

There are no towels.

Could I have an extra pillow please?

Could I have a quieter room?

Would it be possible to change room, it's very noisy?

Checking out

I'd like to pay my bill.

I haven't used anything from the minibar. But I did make one phone call.

I'll be paying by Visa.

The bill should have already been paid by my company.

I think there is a mistake here – I didn't have anything from the bar.

Could I have my passport back?

Can I leave my luggage here and collect it later?

I left you a case this morning.

No it's not that one, it's got a blue stripe on it. Yes, that one.

Could you ring for a taxi for me?

15.12 RESTAURANTS

Formal invitations for dinner

Would you like to have lunch next Friday?

If you are not busy tonight, would you like to ...?

We're organizing a dinner tonight, I was wondering whether you might like to come?

I'd like to invite you to dinner.

Accepting

That's very kind of you. I'd love to come. What time are you meeting?

Thank you, I'd love to.

That sounds great.

What a nice idea.

Responding to an acceptance

Great. OK, well we could meet downstairs in the lobby.

Great. I could pass by your hotel at 7.30 if you like.

Declining

I'm afraid I can't, I'm busy on Friday.

That's very nice of you, but ...

Thanks but I have to make the final touches to my presentation.

No, I'm sorry I'm afraid I can't make it.

Unfortunately, I'm already doing something tomorrow night.

Responding to a non-acceptance

Oh that's a shame, but not to worry.

Oh well, maybe another time.

Arriving at a restaurant

We've booked a table for ten.

Could we sit outside please?

Could we have a table in the corner?

Is there a table free by the window?

Actually we seem to have got here a bit too early.

Are the others on their way?

Would you like something to drink?

Shall we sit down at the bar while we're waiting for a table?

OK, I think we can go to our table now

Menu

Can / May / Could I have the menu please?

Do you have a set menu?

Do you have any local dishes?

Do you have any vegetarian dishes?

Explaining things on the menu and asking for clarification

Shall I explain some of the things on the menu?

Well, basically these are all fish dishes.

Could you tell me what a *name of item on menu* is?

I'd recommend it because it's really tasty and typical of this area of my country.

This is a salad made up of eggs, tuna fish and onions.

Toasting

Cheers.

To your good health.

Making suggestions

Can I get you another drink?

Would you like anything else?

Shall I order (some wine)?

Would you like anything to drink? A glass of wine?

Would you like a little more wine?

Would you prefer sparkling or still water?

What are you going to have?

Are you going to have a starter?

Why don't you try some of this?

Can I tempt you to ...?

Would you like to try some of this? It's called x and is typical of this area.

What would you like for your main course?

Would you like anything for dessert?

The sweets are homemade and are very good.

Saying what you are planning to order

I think I'll just have the starter and then move on to the main course.

I think I'll have fish.

I'd like a small portion of the chocolate cake.

I don't think I'll have any dessert thank you.

Requesting

Could you pass me the water please?

Could I have some butter please?

Do you think I could have some more wine?

Declining

Nothing else thanks.

Actually, I am on a diet.

Actually, I am allergic to nuts.

I've had enough thanks. It was delicious.

Being a host and encouraging guests to start

Do start.

Enjoy your meal.

Enjoy.

Tuck in.

Help yourself to the wine / salad.

Being a guest and commenting on food before beginning to eat

It smells delicious.

It looks really good.

Asking about and making comments on the food

Are you enjoying the fish?

Yes, it's very tasty.

This dish is delicious.

This wine is really good.

Ending the meal

Would you like a coffee, or something stronger?

Would anyone like anything else to eat or drink?

Paying

Could I have the bill please.

I'll get this.

That's very kind of you, but this is on me.

No, I insist on paying. You paid last time.

That's very kind of you.

Do you know if service is included?

Do people generally leave a tip?

Thanking

Thank you so much – it was a delicious meal and a great choice of restaurant.

Thanks very much. If you ever come to Berlin, let me know, there's an excellent restaurant where I would like to take you.

Thank you again, it was a lovely evening.

Replying to thanks

Not at all. It was my pleasure.

Don't mention it.

You're welcome.

15.13   BARS

Suggesting going to the bar / cafe

Shall we go and have a coffee?

Would you like to go and get a coffee?

What about a coffee?

Do you have a coffee machine in the company?

No we usually go to a bar – it's only a few meters away.

Offering drink / food

Can I get you anything?

What can I get you?

Would you like a coffee?

Black or white? How many sugars?

So, what would you like to drink?

Would you like some more tea?

Shall I pour it for you?

Accepting offer

I'll have a coffee please.

I think I'll have an orange juice.

No, nothing for me thanks.

Mistakes with orders

Actually, I asked for an orange juice not an apricot juice.

Questions and answers at the bar / cafe

Do you often come to this bar?

Is there a bathroom here?

Well, I think we'd better get back – the next meeting starts again in ten minutes.

Shall we get back?

Adrian Wallwork

I am the author of over 30 books aimed at helping non-native English speakers to communicate more effectively in English. I have published 13 books with Springer Science and Business Media (the publisher of this book), three Business English coursebooks with Oxford University Press, and also other books for Cambridge University Press, Scholastic, and the BBC.

I teach Business English at several IT companies in Pisa (Italy). I also teach PhD students from around the world how to write and present their work in English. My company, English for Academics, also offers an editing service.

Contacts and Editing Service

Contact me at: adrian.wallwork@gmail.com

Link up with me at:

www.linkedin.com/pub/dir/Adrian/Wallwork

Learn more about my services at:

e4ac.com

A. Wallwork, *Meetings, Negotiations, and Socializing,*   175
Guides to Professional English, DOI 10.1007/978-1-4939-0632-1,
© Springer Science+Business Media New York 2014

Index

This index is by section number, not by page number. Numbers in bold refer to whole chapters. Numbers not in bold refer to sections within a chapter.

A. Wallwork, *Meetings, Negotiations, and Socializing,*
Guides to Professional English, DOI 10.1007/978-1-4939-0632-1,
© Springer Science+Business Media New York 2014